The Origin of Science
and the
Science of its Origin

THE ORIGIN OF SCIENCE
AND THE
SCIENCE OF ITS ORIGIN

by

STANLEY L. JAKI

REGNERY/GATEWAY, INC.
SOUTH BEND, INDIANA

Published by Regnery/Gateway, Inc., Book Publishers,
120 West LaSalle Street, South Bend, Indiana 46601,
by arrangement with the Scottish Academic Press,
Edinburgh.

First published 1979
ISBN: 0-89526-684-9 (clothbound)
ISBN: 0-89526-903-1 (paperback)
Library of Congress Catalog Card No: 78-74439

Manufactured in the United States of America

CONTENTS

FOREWORD

In endowing in 1943 a short course of lectures, Sir Francis Fremantle, a Balliol man from 1891 to 1894, stipulated that the lectures should serve the purpose of making known 'the relation of Science to Christianity and the Christian life (*a*) in essence, (*b*) in its application to the education, thought and life of the masses of our people, (*c*) in its sympathetic explanation of the differences between several religious communities with a view to a common basis for the defence and spread of Our Lord's Gospel'.

Stipulations of this sort are as a rule interpreted with much latitude, a stance which is often a virtue though taking Sir Francis' will at its face value may not necessarily be a vice. A strict compliance with Sir Francis' stipulations is the unabashed aim of these lectures. They aim at showing the crucial role played in the origin of science by a widely shared belief in the first article of Christian creed, an article placing the origin of all in the creative act of God, the Father Almighty, Maker of heaven and earth.

Such an endeavour is far from being novel. One possible way of implementing it does not, however, seem to have been tried so far. It consists in an evaluation of various theories offered about the origin of science and in tracing their obvious inadequacies and inconsistencies to an oversight of or antagonism to that article of faith either in its theoretical relevance or in its historical significance, or both. Authors of those theories had therefore to credit sundry factors—economic, psychological, sociological, political— for the rise of science as a self-sustaining enterprise. That all such factors played a part is all too obvious. That none of them can be taken for the decisive factor should be clear from the inadequacies they display once cast in that crucial role. Their relative, secondary importance could not be treated in the course of five lectures which left room only for a portrayal with quick and broad strokes of the role played by that faith in the origin of science. An adequate science of its origin must therefore include careful attention and appreciation of a faith carried far and wide by messengers of the Gospel.

Also, in accordance with Sir Francis' stipulations, these lectures were geared for an audience not restricted to scholars working in the history of science, my field of study. This demanded a presentation of the topic in a form free from the customary formalities of specialized treatises. Some material supporting the argument is therefore relegated to the notes. The text printed here is closely identical with the one delivered in late April and early May 1977. It is my pleasant duty to express my appreciation to the Master and Fellows of Balliol College for having elected me to the Fremantle Lectureship.

ORIGINALITY AND ORIGINS

Statements on the origin of science are so numerous as to leave little room for originality, but relatively few ideas have so far been offered on what may be called the science of the origin of science. As any other science, this science too ought to be a critical and systematic evaluation of its subject matter. The latter in this case consists of theories and historical accounts of the origin of that very enterprise, science, which gives to modern times the label scientific, a label that cannot be attached to any previous age and seems to have little connection if any with historic Christianity. Undoubtedly, we have come a long way from that virulent anti-Christian tone which dominates Condorcet's protrayal of intellectual history and had set the tone of most cultural histories written during the nineteenth century and beyond, histories in which the modern origin of science plays a pivotal role. Yet, even if stark antagonism to Christianity no longer assures academic laurels in an automatic way as it did in the not so remote past, a studied slighting of the Gospel as a cultural factor, let alone as a factor in the rise and progress of science, is easily taken today as a partial assurance of intellectual respectability and a hallmark of scholarship.

On the face of it no such slighting is implied when the birth of science is assigned to a place and time, such as classical Greece, long antedating Christianity. The phrase—Greece, the birthplace of science—has been presented to modern man's eyes and dinned into his ears so often and so systematically that, even if he is a historian of science, he may not have second thoughts about it. That second thoughts are in order can be seen even from the words of Alexandre Koyré who placed the birth of science in ancient Greece as if such a placing were above any doubt. His words are all the more worth recalling as they were addressed in 1961 to a large and impressive gathering of historians of science in Oxford.[1] Although disagreement is usually the rule in such occasions, no exception was taken to his statement on Greece as the birthplace of science. According to Koyré 'we can adequately explain why science was not born and did not develop in Persia and China (the great bureaucracies, as Mr. Needham has pointed out, are everywhere hostile to independent scientific thought) and although we may be able to explain why it was possible for it to be born and to develop in Greece, we still cannot explain why it did so in fact'.[2]

The first part of Koyré's statement had for its principal aim the defence of his and of his followers' programme, an opposition to the trend which specifies the source of science in practical and social needs. The second part, or the inability to say why science did arise in Greece, reveals something of the Platonism of Koyré, who did not endorse Platonist ontology in his writings on the history and philosophy of science. His basic explanatory framework seems to have been closer to a psychologism based on the doctrine of Gestalts which he not only had found in the writings of Gaston Bachelard, but found it applied there to the mental 'mutations' that constitute scientific revolutions.[3] Nobody would, of course, question the fact that a revolution in thinking occurred during the period which constituted Koyré's principal field of study, the period stretching from Copernicus to Newton, and more specifically pivoted on Galileo. It is, however, regrettable that historians of science by and large failed to challenge Koyré on a rather obvious and unique fact of the Galilean revolution, or perhaps birth of science. Their challenge should not have touched on the measure to which Galileo was indebted in some particulars to his late-medieval predecessors, a measure which Koyré invariably sought to minimize. Their challenge should have rather aimed at the Galilean birth of science in so far as it was a true birth. The Galilean birth of science was for Koyré, and for many others since Kant, a quick birth, a sudden shift in mental vision. However that may be, in so far as it is a birth, it should appear, in retrospect at least, obviously different from the kind of birth science had in classical Greece. No dispute would in all likelihood follow the statement that as the seventeenth century approached its mid-point, the scientific enterprise had reached a stage where its vigorous continuation had become assured. This is in sharp contrast to what happened to science in classical Greece. The birth science had there was not a viable birth. By the time Euclid produced the most memorable monument of Greek science, Greek thought had been caught in a circle out of which neither the Stoics nor the Epicureans could provide that escape route which the Platonists and the Aristotelians had already blocked. Nor could for that matter an Archimedes, although the avid reading of his works in late medieval times acted, according to Koyré, as a midwife for the Galilean birth of science.

All this can easily be seen from the retrospective vantage point of twentieth-century scholarship. But the historian, or the historian of science to be specific, is expected to offer more than a retrospect, reliable and informative as it may be. He also has to make an effort to trace out the history of his retrospect. He should ask himself in particular whether the retrospect was already consciously achieved at its logically earliest phase. The phase in this case is the early part of the seventeenth century which in the eyes of many witnessed the first viable birth of science. That the period in question may indeed be rich in evidencing a strong awareness of this retrospect can be suspected as soon as one makes the tempting association of viable birth with Francis Bacon's famous characterization of his age as 'the masculine birth of time'.[4] Bacon also spoke of himself as the trumpeter of new times,[5] and the bell-ringer of a new age,[6]—so many uncanny portrayals of his robust awareness of the fact that something very new was in the making. In fact, he felt he had provided the instrument ushering in the new golden age of applied science. It never dawned on him that his method, which he compared to an unfailing machine, was applying science before having it on hand. The other towering figures of the age, Descartes, Kepler, and Galileo, matched Bacon's ambitious *New Organon*[7] with no less ambitious titles which fully conveyed their authors' awareness of being innovators. Descartes made his bow with a discourse on *the* method, Kepler with a launching or *prodromus* of a series of works on the cosmos, and Galileo with nothing less than a messenger from the stars.[8]

Unlike the name of Bacon, the names of Descartes, Kepler, and Galileo are remembered as the names of great scientific discoverers. If Bacon promoted science he did not do so by making scientific discoveries. Yet for all his failure to make even a minor discovery, Bacon remains an all-important figure in that period that witnessed the viable or masculine birth or origin of science. It is well to remember that the *New Organon* was written two decades or so before Descartes published his *Géométrie* and Galileo his *Dialogue on Two New Sciences* and before Jeremiah Horrocks saved Kepler's true feats for Newton and science. A sensitive antenna of a scientific future Bacon certainly was, and as such he repays the effort of going back to him. But going

back to Bacon is imperative for another reason which is
better seen with Galileo as a background. In his famed
Dialogue on the Two Chief World Systems Galileo proposed
'to show to foreign nations that as much is understood of
this matter [of the earth's motion] in Italy, and particularly
in Rome, as transalpine diligence can ever have imagined'.
Although never generous with praise for his forerunners, it
was natural for Galileo to see himself as riding on the crest
of a wave, the renaissance of arts, sciences, and letters that
had made Italy by then the envy of the rest of Europe. The
centre of that Italy was Rome, then as now, and there was
more than clever tactic in Galileo's words that his *Dialogue*
was to show not only that 'everything was brought before
the attention of the Roman censorship' but also that 'there
proceed from this clime [of Rome and Italy] not only dogmas
for the welfare of the soul, but ingenious discoveries for the
delight of the mind as well'.[9]

Whatever of Galileo's devotion to the Church, and whatever of his good intentions of staving off a potential conflict
between dogma and discovery, the climate of post-Galilean
Italy was set more by dogmas than by discoveries. While it
was true that from his *Dialogue* the rest of Europe could
perceive that 'if other nations navigated more, we [the
Italians] have not theorized less',[10] the rest of Europe could
see before long that as other nations engaged in navigation
more and more, the Italians did less and less theory, or
science. A generation or so after Galileo's death the centre
of scientific learning was no longer Italy. The new centre
was not France either, mesmerized by Descartes, on whom
d'Alembert laid the blame almost exactly a hundred years
after Galileo's death for the stagnation of French science.[11]
Science appeared to be a very minor issue indeed when in
the closing decades of the century the French were, through

the works of Fontenelle and Perrault, suddenly confronted
with the quarrel concerning the respective merits of ancient
and modern learning.[12]

That the same quarrel had already been raging in England
for some time and with strong emphasis on science indicates
that the centre of science was England by the 1660s. Although
this is all too clear in retrospect, the wisdom of hindsight
should once more be matched with evidences from the contemporary scene. That England could boast of a Harvey
and a Boyle was, of course, no small matter; nor should one
underestimate the significance of Hooke's *Micrographia*, just
fresh from the press. But when it came to single scientific
feats, the France of those times was not lacking them, nor
were Italy and Germany suddenly deprived of outstanding
scientific merit following Galileo's and Kepler's departure
from the scene. Pascal and Fermat vouched for France, Tor-
ricelli for Italy, Guericke and Leibniz for Germany. But
unlike Germany, France, and Italy, England was rapidly
enveloped in a climate of opinion, boisterously joyful over
the potentialities of the newly found scientific approach to
things and problems. The exuberance related to more than
matters scientific. In no other place but in England could the
mid-seventeenth-century man achieve that measure of self-
esteem which prompted Milton to state: 'God is decreeing
to begin some new and great period in his Church, ev'n to
the reforming of Reformation itself: what does he then but
reveal Himself to his servants, and as his manner is, first to
his English-men'.[13] Only a special climate of opinion could
be so appreciative of one's own place and time.

As to the phrase, 'climates of opinion', it was possibly the
only lasting contribution of Joseph Glanvill,[14] a convert to
the newly founded Royal Society, membership in which was
at that time still a calculated risk, royal patronage notwith-

standing. Boyle, a scientifically respectable figure among the amateurs and dilettants that founded the Royal Society, was urged and repeatedly so by its critics to break ranks with it. But Boyle was not only captive of the new climate of which the Royal Society was the chief breeding place. He was also a most pervasive propagandist of that climate. Like other members of the Royal Society resolved to swear *nullius in verba*,[15] he was ready to take as Gospel truth Bacon's words on the new experimental method.[16] Although generally cautious in stating his scientific conclusions, Boyle would have hardly disagreed with the soaring prose of Henry Power, his fellow member in the Society, who concluded his rather primitive account of new experiments 'microscopical, mercurial, and magnetical', by declaring that his times were the age wherein 'Philosophy comes in with a Spring tide' and the age that was bound 'to lay a new Foundation of a more magnificent Philosophy, never to be overthrown'.[17] Nor would Boyle have disagreed with Simon Patrick, as ardent a defender of the experimental method as he was innocent of it, who claimed that it was as impossible to stop the new philosophy as 'to hinder the Sun from rising, or being up, from filling the whole Horizon with light'.[18] Boyle himself was eager to support by reasoning as well as by experimenting 'the inquisitive genius of latter years' about which Patrick wrote that 'like mighty wind [it] hath brushed down all the Schoolmen's Cobwebs'.[19] Boyle certainly was one of those who felt, again to use Patrick's words, that 'there is an infinite desire of knowledge broke forth in the world and men may as well hope to stop the tide, or bind the Ocean with Chains, as hinder free Philosophy from overflowing'.[20]

Spring-tide, mighty wind, the sudden breaking forth of light over the whole horizon, the feeling of being seized with an infinite desire of knowledge, are certainly indicative of something very new in the air. With the exception of Newton, members of the Royal Society were unanimous in crediting Bacon as the main source and chief articulator of the novel spirit by which they were seized. William Petty called Bacon 'the Master-Builder',[21] John Wilkins described him as 'our English Aristotle',[22] and Boyle spoke of him as 'the first and greatest experimental philosopher of our age', and as 'the great architect of experimental philosophy'.[23] Exhortation on doing science Bacon's writings certainly contained, but science little if at all. Boyle could not have been further from the truth in seeing his experiments described in his *Certain Philosophical Essays* as a continuation of Bacon's *Sylva sylvarum*.[24] In devising and interpreting his experiments Boyle heavily relied on ideas, those very ideas and theorizing on which Bacon frowned and which were conspicuously absent from his writings. Bacon was a jurist at heart, the explanation of the truth of Harvey's biting remark to the antiquarian, John Aubrey, that Bacon wrote philosophy like a Lord Chancellor.[25] As a jurist of natural philosophy he was interested in factuality, not in explanation and self-reflection.

Thus the admirers of Bacon in the Royal Society and elsewhere could find in his writings little if any insight into the reasons and origins of that novel outlook that seized them. Bacon himself ascribed to a mere accident of the times[26] the occurrence to him of those very novel ideas that led him to formulate that new tool or organon, the application of which automatically assured, so he believed, the masculine birth of times.[27] As to the so-called times, they are in a sense the fiction of the historian, especially when he personifies ages, epochs, centuries, and decades, and transfers to them the ability of that thinking which only individual human beings do possess. As to the accident or chance, it is the confession of our ignorance which can be of two kinds, a practically insuperable ignorance or one about which something can and ought to be done.

It is the latter kind of ignorance which Bacon had in mind and he chose to acquiesce in it. This would not have happened had he not declared the study of miscarriages or stillbirths of time uninstructive.[28] He might even have perceived a most momentous truth about his method of experimental philosophy. The truth concerned the 'times' giving birth to his method. Pondering over the miscarriages of time or science might have given him second thoughts on imagining his method coming forth full-grown and fully armed from the womb of his times. Clearly, if Bacon had perceived that science had already made several efforts to be born, though in vain, its birth in his times might not have appeared a mere accident of time. But Bacon saw little in past history resembling a parturition, however unsuccessful, of science. All intellectual history appeared to him barren except three periods each lasting less than two hundred years. The three periods belonged to ancient Greek and Roman times and to the most recent times, that is, as Bacon put it, to us, 'the nations of Western Europe'.[29] The Arabs and the scholastics were written off by him in a line, but so were most Greeks as far as science proper was concerned. Owing to the influence of Socrates, natural philosophy fell into disrepute among the Greeks after flourishing, to recall Bacon's words, for a 'brief particle of time'.[30]

While Bacon was right concerning the impact of Socrates, he was wholly wrong in slighting Greek science. This is why he could not see there a major stillbirth or miscarriage of science. Greek science appeared to Bacon, the grim empiricist, too theoretical to deserve consideration, although he took the view that practically all the science of his time had come from the Greeks.[31] His empiricism had greater sympathy for the allegedly inventive Egyptians, a sympathy which is less revealing of the Egyptians than of a Bacon never able to do justice to reason. The sympathy was a stark revelation of the blindfolds empiricism was capable of putting on one's mental vision. According to Bacon, among the Egyptians, who 'rewarded inventors with divine honors and sacred rites, there were more images of brutes than of men; inasmuch as brutes by their natural instincts have produced many discoveries, whereas men by discussion and the conclusions of reason have given birth to few or none'.[32] Bacon failed to suspect the irony that the names of those inventors were never recorded, most likely because in Egyptian context the bestowing of divine honours often meant the replacing of human heads with those of animals and these stood for types, not for individuals who alone can produce theories and inventions.

A more sympathetic view of theoretical reason coupled with pondering the miscarriages of time might have helped Bacon notice that the times which he wanted to bring about had already been in the making for some centuries. This, of course, would have become a shocking discovery for him, because it would have presented him with the necessity of a 'reforming of Reformation itself', to recall once more Milton's penetrating remark. Such a prospect, even if taken as a purely historical one, should seem rather connatural with Sir Francis Fremantle's wish that the lectures founded by him should offer, as was already noted, a 'sympathetic explanation of the differences between the several religious communities (denominations, as we would say nowadays) with a view to a common basis for the defence and spread of our Lord's Gospel'.

In this age of ours, which is happily far more ecumenical than former times, it will not perhaps sound polemical to state that Bacon and others proud of the Reformation went too far in depicting the ten long medieval centuries as deprived of light, spiritual and philosophical. In doing so they deprived themselves of an insight, both spiritual and philosophical, which would have saved them from looking upon themselves as the lucky accident of times, a view hardly philosophical or spiritual. From the spiritual viewpoint 'the nations of Western Europe' had been not for two but for several centuries distinctly different from the Greeks of old. The latter, as Bacon aptly remarked, failed in their science, because under the impact of Plato and Aristotle, they made nature replete with final causes to such an extent as to feel no need of God,[33] the true source of purpose or finality. Whatever the popularity of Aristotelian science during the Middle Ages, the medieval nations of Western Europe did not follow Aristotle into pantheism, precisely because their Christian faith kept alive in them a vivid need of God. The same faith also provided them with a philosophical insight, namely, the insight about the contingency of the world.

The contingency of the world derives, of course, from its createdness, a point most natural for a Christian to entertain and a point most naturally espoused during those Middle Ages which saw Christianity become for the first time a broadly shared cultural matrix. As a philosopher Bacon was not wholly unoriginal in noting the logical basis which is provided for empirical science by the Christian dogma about the createdness or contingency of the world. As a historian Bacon could have achieved striking originality, had his strong prejudices against medievals and Middle Ages not prevented him from spotting the early origins of science in the High Middle Ages, an age steeped in a view about the contingent origin of all which only the Christian creed could provide. But Bacon and his admirers in the Royal Society recalled the medieval past with no hint to the unrelenting resolve with which the strictly Aristotelian Averroists had been opposed by the best medieval Christian philosophers. Their fight was not about subtleties of logic, the sole domain where the medievals were given intellectual credit by Bacon and the protagonists of the Royal Society. The fight in question was about the nature of reality and knowledge, crucial issues for Christian or supernatural faith as well as for natural science. Had the medievals not turned the first article of the Creed into a climate of opinion, Copernicus could have hardly mustered courage for working out in full the bold vision of heliocentrism. Had Copernicus' aim been a Baconian usefulness he would not have needed the courage to propose a stark innovation. The usefulness which navigation and calendar-making could derive from heliocentrism, as articulated in Copernicus' *De revolutionibus*, did not exceed the corresponding benefits that had already been gained from the geocentrism of Ptolemy's *Almagest*. The true appeal of Copernicus' system lay in its being a marvellously simple idea about the world, the kind of idea which for Copernicus pointed to the Creator and about which Bacon later was to say that such ideas were the stamps of the Creator himself,[34] but which strangely enough had no proper function in his experimental method.

Ideas had no appeal to Bacon and therefore he could not be sympathetic to the history of ideas including that of Christian Aristotelianism and of Christian Platonism, both of which, and obviously the latter, played a crucial role in that viable birth which Copernicus, Kepler, and Galileo secured for science. As in other cases, in Bacon's case too, a strong narrowing of the mental horizon provided the trap from which escape was sought though in vain. The one fighting Aristotelian preoccupation with notions and natures argued that it was useless to speak with Copernicus about the rotation and orbiting of the earth and planets until the 'nature' of rotatory and orbital motions had been fully clarified,[35] a task as Aristotelian as was that Aristotelian ether whose nature it was to move in a circle. While Bacon's critique of Copernicus was pre-empted by the inner logic of his empiricist method forbidding questions about nature and natures, it was the empirical facts that best refuted Bacon's claim that there had been a steady progress of practical inventions.[36]

Had there been such a progress, a progress clearly implied in Bacon's grossly empirical method, ancient history would have certainly been different, a point sufficiently clear even from that rudimentary stage at which the study of the history of technology stood in Bacon's time. Bacon himself relied on that study as he remarked that Greek learning resembled the intelligence of boys who could talk but not generate.[37] The simile was marvellous, giving as it did a tantalizing glimpse of a stage in which learning (science included) could not generate its further progress. But Bacon did not consider why the allegedly steady progress of technical inventions failed to carry the Greeks of old beyond that 'boyish' stage, or more truthfully, why their science proved to be a miscarriage or stillbirth of time.

It would not be altogether justified to press Bacon and the Baconians for not having pondered a few particulars, even if they had been aware of them, such as the failure of Egyptians of old to devise pulleys and of the Maya to make wheeled vehicles, tools that certainly would have been useful for their monumental constructions. But Bacon and the Baconians should have certainly reflected on the medieval cathedrals gracing everywhere their countryside. Those cathedrals embodied powerful and, with respect to previous ages, strikingly novel architectural techniques that were hardly prompted by utilitarian considerations. Glanvill and other protagonists of the Baconian programme of learning kept silent about those cathedrals, products of an age which they credited only with subtlety in logic. Their silence was all the more curious because in a strikingly technological sense those cathedrals were not at all silent. From their façades the hours were struck far and wide by weight-driven clocks, one of the most crucial inventions of all times and an invention which had found industrial application already in medieval times. For Bacon and the Baconians it would have been inconceivable that four hundred years after them historical scholarship would locate one of the most important turning points in the history of technology, Bacon's archetype of science, in the Middle Ages and trace there that turning point to that very new mentality which Bacon felt he had first articulated and in contradistinction to medieval mentality.[38] At any rate, that the Romans, practical engineers as they were, failed to invent the weight-driven clock was difficult to explain on Baconian principles, though perhaps not as difficult as to account, on the basis of the same principles, for the persistent inertia of the Chinese with printing, with magnets, and with gunpowder. Did not Bacon hail these three as the very inventions which led man to the threshold of the masculine birth of time?

Clearly, it would have made no sense to say that the Chinese had no need of movable types, the kind of printing developed by Gutenberg, though hardly without the help of some late-medieval exploratory work, or that the Chinese had no need of a compass, that is, a navigational instrument far superior to a loadstone. As to the gunpowder, to say nothing of guns, the Chinese learned a great deal from Jesuit missionaries. The usefulness of warfare with explosives would not today be admitted by all, but Bacon and the Baconians had no doubts on that score. A peacefully utilitarian use of explosives was still centuries away when Glanvill spoke in glowing terms of gunpowder as that 'other great German invention'.[39] He chose not to elaborate on the broader medieval background of that same invention which he ascribed to Bernard Schwartz, not only a German but also a monk.

Equally perplexing should seem the praises heaped by some Baconians on Descartes. It is doubtful that almost thirty years after Descartes had slighted Bacon's method in a letter to Mersenne,[40] the Baconians had not yet learned of it, as Mersenne was well known for his rapid transmission of informations received by him. But feeling confident about possessing the philosopher's stone, the Baconians readily claimed for themselves the good points of even their most resolute antagonists. In describing Descartes as 'one of the greatest Wits ever the Sun saw, a Person too great for praise, designed by Heaven for the Instruction of the Learned World',[41] Glanvill was probably thinking of analytical geometry and perhaps even more of Descartes' very Baconian aim to provide the means, however un-Baconian, of turning man into 'master and possessor of nature'.[42] The phrase, which soon became taken for the epitome of the modern mind, was already five centuries old when Descartes pulled it out of his subconscious which had been nourished by the catechism and the deeply scholastic tone of instruction he had received from his Jesuit teachers at La Flèche. To become master and possessor of nature was a primary injunction made on man at the very outset of the history of salvation which through the preaching of the

Gospel became part of a widely based tradition and the leaven of a cultural matrix that turned medieval Europe into its specific and historic identity. In that tradition, inherited by Baconians and Cartesians alike, the sense of a God-given duty to dominate nature was accompanied by respect for ancient learning. The latter received its pregnant formulation—'we see farther because we stand on the shoulders of giants'—from Bernardus Sylvestris[43] almost exactly at the time when Hugo of St. Victor wrote in his *De sacramentis* that 'God put man on earth to make him the master and possessor of nature'.[44] It shows something of the intellectual turbulence of Bacon's time that the phrase of Bernardus Sylvestris had become a shibboleth in an acrimonious controversy concerning the respective merits of ancient and modern learning, whereas Hugo of St. Victor's anticipation of Descartes' motto was conveniently overlooked.

Illogical as this could be, it matched the blurred view offered by Bacon and Baconians about the origin of science. As far as useful inventions were concerned there was a sudden growth of them in the centuries immediately preceding Bacon and his generation. This demanded a better explanation than Bacon's invoking the peculiarity of the times. There was no more explanation in falling back, as Sprat, Glanvill, and others did, on the genius of the times.[45] Overawed as they were by the empiricist method, Sprat and Glanvill could find in the word genius hardly more than a mere word and not a lead to deeper understanding. Even if genius were equal to curiosity, one could still ask why certain times had an unusual share of it. At times Sprat and Glanvill attributed this to the reaction which should set in when everything seems to go wrong. Obviously, it was easy to see by the seventeenth century that many things were wrong with Aristotle's physics and cosmology. But awareness of the defects of Aristotelian physics was considerably older,[46] though it was not a necessary prerequisite of progress. Copernicus, for one, supported the physical truth of his heliocentric cosmology with recourse to notions germane to Aristotelian physics.

Recourse to the genius of the times should seem therefore a facile exercise of writers too imbued with their sense of originality. The result was not curiosity about origins but a curious disdain for the importance of clarifying it. Sprat, who specified insatiable curiosity as the temper of mind of the new age, brushed aside the question which was in a sense the question of the viable birth of science. In speaking of the recovery of all ancient knowledge during the preceding couple of generations he refused to single out the most important among the three reasons he listed for that recovery: They were the spread of knowledge through printing, the hatred 'against the blindness and stupidity of the Roman Fryers', and the Reformation which put men 'upon a stricter inquiry into the Truth of things'. Whatever the cause was, he declared, 'I will not take much pains to determine'.[47] Apparently, all three reasons were equally important.

This cavalier approach was a piece of that self-satisfaction which characterized the attitude of the protagonists of the new learning toward the old times invariably identified with Aristotelianism. The dispute between ancient and modern learning,[48] which in itself would have been most beneficial for a clarification of the origin of originality right then when originality blossomed full, could therefore hardly produce more than hollow declarations of victory coupled with the abuse of the opponent. For Glanvill the Aristotelians distinguished themselves by stubborn reluctance to consider the facts. Indeed, what he said of his dispute with one of

them suggested that two generations after the construction
of the telescope the genuine Aristotelian still refused to look
through it on the ground that the 'Glasses were all deceitful
and fallacious'. 'Which answer [of the Reverend Disputer]
minds me', Glanvill added, 'of the good Woman, who when
her husband urged in an occasion of difference (I saw it, and
shall I not believe my own eyes?) replied briskly, Will you
believe your own eyes, before your own dear Wife?'[49]

Obviously then if the Aristotelian defenders of ancient
learning were such as to deny to others the right to believe
their eyes, they had no right to have their contributions
considered in those sketchy histories of science which both
Sprat and Glanvill made part of their defence of modern
learning. In listing the names of ancient Greek mathematicians, Glanvill pointedly remarked that Pappus did not
mention Aristotle as a mathematician which he certainly
was not.[50] But in the Baconian perspective of natural
histories no small credit should have been given to that
Aristotle about whom two centuries later Darwin was
prompted to say that compared with him Linnaeus and
Cuvier were mere children.[51] But the merits of Aristotle, the
biologist, could but come to naught when the circulation
of the blood came to be taken for the first meaningful page
in biology. Clearly, a history of science sketched with the
purpose of vindicating the empiricists equated with anti-
Aristotelians could not amount to a history respectful of
facts, big or small. Among the big facts dismissed with a
bare word or two was the fate of science in ancient cultures.
Sprat mentioned only the Egyptians and only to remark that
they had 'in good part invented the learning of the East'.[52]
This was a gross superficiality even for Sprat's time. Within
its confines one could not even ask the obvious, namely, why
the Egyptians had settled with a very primitive form of

writing, reckoning, and measuring. The question of cultural miscarriage had, of course, no room in the Baconian perspective of steady progress. Had Sprat asked the question about science in Ancient Egypt, he might have offered something similar to what he said in connection with the Arabs. He saw them as victims of the 'barbarous Turks'.[53] Glanvill was more enlightening as he stated that the Arabs did not improve on Greek learning and science, though he left it at that.[54]

Curiously, neither Sprat nor Glanvill pondered why the Egyptians, Babylonians, and Arabs were not more curious about those very practical needs that burdened man everywhere and all time. Neither Sprat nor Glanville referred to the Chinese, an omission that should seem glaring even from the vantage point of the 1660s. By then at least half a dozen reports on China had been circulating widely in Europe. They all were written by Jesuit missionaries starting with Ivan Gonzales de Mendoza's *Histoire* in 1588,[55] followed by Nicholas Longobardi's *Nouveaux Advis* (1602)[56] and by Nicholas Trigault's *De christiana expeditione* (1615),[57] to mention only the major ones. Then a generation later Alonso Semedo published his *Histoire universelle du Royaume de la Chine* (1645),[58] Martin Martini his *Decas prima* (1658)[59] and Melchisedech Thévenot his *Relations* (1663–72), a collection of unpublished China reports containing Prospero Intorcetta's translation of a book of Confucius under the title, *Sinarum scientia*.[60] From all these books one could find that in precisely the fields of crafts and science the Chinese were bafflingly behind. Longobardi gave in 1602 the not-at-all unreliable evaluation that Chinese learning stood in 1600 where European learning was at the time of Cicero.[61] Another common trait in these China reports, curiously neglected by Needham,[62] was the reference to European

superiority which their authors traced to the pronounced
curiosity of the European mind.

There was, of course, much defect in all these China
reports. Whatever their methods, the Jesuits, in line with
their calling, wanted to serve in China neither Europe, nor
science, but the Gospel. That they failed to perceive, as did
the Protestant Baconians back in Europe, the connection
between their curiosity and their being steeped in some
Gospel truths, should seem curious though not surprising.
Although nothing is more needed for making a discourse
than the air we breathe, nothing is proportionately less reflected upon than the air itself, be it the climate of opinion
or the temper of the age, religious or secular. To be sure,
Longobardi wrote that nothing amazed him and his mission-
ary colleagues more than the fact 'that such enlightened
people are so blind about things that are fully clear to us'.
Those clear things were, according to Longobardi, 'that
there is one God, Creator and Governor of the universe,
that the rational soul is immortal and it has to be rewarded
or punished according to his merits and the like'.[63]

That Longobardi believed already in 1602 that he had
found these truths expressed in very ancient Chinese writings
and paintings was the first sign of what later became the
Rites controversy which produced so much grist for the mill
of resolute scoffers at the Gospel such as Voltaire and company. As everybody knows today, Longobardi and his sinophile colleagues read too much into Chinese tradition.
Equally reprehensibly, they failed to read in depth their own
tradition of evident truths. That the world was ruled by the
Creator and similar tenets were evident to them was due to
their being steeped in the Gospel. Their European curiosity
about the world also had much to do with their and their
forebears' steady exposure to the same Gospel over many

centuries. European originality, which most palpably evidenced itself in science, had its origin in the Gospel, the preaching of which planted deep in European minds, long before Bacon and Descartes, the conviction that the universe was the rational product of the Creator and that as Christians they had to become masters and possessors of nature. The new organon of science was not in the voluminous fumbling of Bacon with mostly irrelevant facts but in the conviction shared long before him of the fact that since the world was rational it could be comprehended by the human mind, but as the product of the Creator it could not be derived from the mind of man, a creature. Of course, the seeds of the Gospel, or more concretely of Christian creed, which claims both the Old and the New Covenant, needed a natural matrix, social and psychological, to issue in palpable originality and genius about nature. When the Jesuits blamed Chinese autocracy and bureaucracy and the not-so-curious Chinese mind for the backwardness of science in China, they were in a sense as modern as those twentieth-century historians of science who think it is enough to refer to social conditions and to psychological structure in order to cope with the question of the origin of science and with the originality of those originating it. In doing so the Jesuits not only anticipated twentieth-century modernity but also its superficiality. They deserve less criticism on that score than Koyré and other historians of science. The latter could have easily noticed the stultifying results in studies concerned with the origin of science whenever that origin had been sought in studied neglect, let alone in plain contempt of the Gospel. That there is indeed something stultifying in such studies may be anticipated from the mere title of the next lecture: From Paris to Mongolia.

FROM PARIS TO MONGOLIA

There are books which seem to everybody all-important at the time of their publication and books that show their true importance only some time afterwards. A chief among the former was, in the closing years of the seventeenth century, the *Nouveaux Mémoires sur l'état présent de la Chine* by Father Louis Lecomte of the Society of Jesus and mathematician to his Majesty, Louis XIV. The two-volume work,[1] composed of fourteen letters addressed to various French potentates, ecclesiastical and civic, went between 1696 and 1701 through five editions, was quickly translated into English, German, and Dutch, and was condemned in due course by the theologians of the Sorbonne and afterwards put on the Roman Index. Written in an engaging style, graced with numerous illustrations, printed in a handy format, Father Lecomte's book covered many topics, but its real aim was to show that Confucius still professed monotheism and moreover that monotheism was the generally shared belief in pre-Confucian China.[2]

Father Lecomte offered no explanation, or at least no convincing one, of how the Chinese of old could for many generations remain immune to the effects of original sin. A long-standing attachment to belief in one God was hardly a feat befitting fallen man which the Chinese were no less than were the other offspring of Adam. Instead of offering a theological explanation to a theological problem, Father Lecomte took to the modern way and put the problem into a cultural perspective. By the time his reader reached the *matière célèbre* in the book, he was already regaled with a rich vision of

Chinese geography, politics, history, literature, arts, crafts, and last but not least science. In presenting all these matters Father Lecomte was fully aware that he was a European writing to Europeans about a vast but distinctly non-European culture.[3] Some of the differences must be a source of amusement now as then. On being told that in Europe a king can be succeeded by a queen the Chinese replied 'laughingly that Europe is the Kingdom of women'.[4]

Beneath such trivial differences there lurked depths, but Father Lecomte seemed to be intent on skirting deep waters. Time and again he found the Chinese culturally backward as compared with contemporary Europe but the most he offered in the way of explanation was that the Chinese 'did not receive the spirit of penetration and subtlety so necessary for those who apply themselves to the knowledge of nature'.[5] If this was true, then the Europeans' skill and subtlety with nature was also a special gift of God, a conclusion which was not drawn by Father Lecomte. Logic did not seem to be his forte, but it was a weak point in his view of science that prevented him from seeing a connection between Gospel and science in that European culture which was clearly permeated by both.

The European science which Father Lecomte and his Jesuit confrères had in mind was Euclid and Ptolemy with some new inventions, such as the telescope, thrown in for good measure. The Jesuits in China kept silent on Copernicus and his defence by Galileo. Father Lecomte could hardly suspect that it was largely because of a book, Newton's *Principia* (already for three years in print before he had set sail for China), that the Chinese would, as he reported their reaction, 'take forever the Europeans for their masters in these matters'.[6] Father Lecomte should not, of course, be taken to task for not speaking of Newton while in China. The most he could have learned about the *Principia* was a summary evaluation by its first French reviewer, who being Cartesian took the view that it contained much good mechanics, but no physics at all.[7] A telling misapprehension which showed that the *Principia* was one of those few books that reveal their importance only some time after their publication. Even a Leibniz, so much abreast of the latest in science and philosophy, had become familiar with the *Principia* only two years after its publication.[8] In fact as late as 1730 when Pope coined his famed epigram—Let Newton be, and all was light—the light in question was still largely confined to England and even there in a rather qualified way. By throwing his weight around, Newton the scientist-turned-administrator established not so much the *Principia* as a 'Principate'which consisted of his friends, admirers, disciples, and protégés receiving the best academic posts and benefices. But all that academic coterie of Newton paid relatively little attention to the *Principia* or more specifically to the development of its immense potentialities. David Gregory's *Astronomia* published in 1702 was certainly a useful (and the first) textbook-presentation of Newton's celestial dynamics but it advanced not a whit the science of physics. The same was also true of Maclaurin's *Account of Sir Isaac Newton's Philosophical Discoveries* published in 1748.

It was almost exactly at the halfway point between these two publications that Swift came out in 1726 with *Gulliver's Travels*.[9] In this age of ours eager to cultivate the history of science Swift's classic is not only an entertainment for youngsters, but also for historians of science. They find there not only a specification of two satellites for Mars one-hundred-fifty-one years before their discovery,[10] but also an acid lampooning of the Royal Society. Concerning those satellites of Mars, Swift's was blind luck, but perhaps even as an outsider to science he sensed something about the lack of creativity of the Society's Fellows as he described the Academy of Projectors of Lagado. The creativity by which Newton's *Principia* defied the empiricist precepts (Newton hardly ever referred to Bacon) yielded to that experimenting which easily degenerates into tinkering. In the process one would, and this was a principal point made by Swift, become engrossed so much with things as to become contemptuous of ideas and of the very words that carried them. In debasing biblical history to a pretentious game with chronology Newton produced a classic example of how a mind most creative in one respect can become pedestrian in others. A chief glory of the Academy of Projectors of Lagado was a machine that produced books,[11] and one wonders whether Swift had in mind the *Theologiae christianae principia mathematica* by John Craig, professor of mathematics at Cambridge. The quarto booklet contained a quantitative evaluation of the decrease and ultimate vanishing of credence given to a historical fact (Jesus' life) as time went on. The result was the specification of the year 3150 A.D. as the end of the world marked by Jesus' second coming, that is, by the fulfilment of His question, 'when the Son of Man cometh, shall He find faith on the earth?', recorded by Luke.[12] Craig's handling of it would have done much credit to the mechanical philosophers of Lagado.

One of the projects of the Academy of Lagado was the replacement of spoken words with kits of things appropriate to one's own field of interest.[13] One advantage of implementing the project would have been a decrease of strain on one's lungs, a benefit that may appear of some merit in this age of lung cancer, generally attributed to smoking and never to excessive talking. At any rate, useful as the kit of things replacing words could appear, it was not introduced because, happily for mankind, 'women, in conjunction with the vulgar illiterate . . . threatened to raise a rebellion, unless they might be allowed to speak with their tongues'.[14] The remark, 'such constant irreconcilable enemies to science are the common people', was added by Swift with tongue in cheek and with a prophetic touch. Within a few years young David Hume was to retire to France for the purpose of drawing out the full logic of Locke's empiricism through a resolute elimination of all unreliable ideas. The result was that all ideas became suspect to him and in line with this he urged in the concluding paragraph of his *Enquiry concerning Human Understanding* the committing to flames of all books that contained no quantities and matters of fact. Such was in Hume's radical empiricism or sheer sensationism the beginning of the ultimate elimination of the common sense of all people, common or not, through the discarding of many a word in favour of sensuous objects, or rather of mere sensations, a programme akin to the one advocated in the Academy of Projectors of Lagado.

The growth of empiricism from Locke to Hume was an English phenomenon paralleling the decline of science in England. It revealed once more the true character of Bacon's empiricism, a method rejected tellingly enough by the few whose work formed exception to that decline. Thus John Hunter, who raised surgery from the crudities of army barbers to the level of science, noted that without the understanding of the principles of his art, the surgeon would 'resemble the Chinese philosopher whose knowledge consisted only in facts'. That there was no science to be had on such a basis was the gist of the contrast drawn by Hunter: 'In Europe, philosophers reason from principles and thus account for facts before they arise.'[15] Hunter's remarks may have been provoked by Hume's lengthy essay, 'Of the Rise and Progress of Arts and Sciences', in which Humean empiricism proved itself to be no match to the unique birth of science and especially to its stillbirth in China, to say nothing of other ancient cultures.[16] Hume was not the first and last empiricist to take refuge in this connection in words about genius that proved indeed to be mere words. That Hume, who did not write elsewhere about China, paid attention in that mostly forgotten though very revealing essay to the failure of the Chinese of old to give rise to science was in all likelihood due to an external stimulus. It might have consisted in his reading three long letters of Father Parrenin to de Mairan which appeared in the *Lettres édifiantes* between 1734 and 1743.[17] The correspondence between de Mairan, future perpetual secretary of the French Academy, and Parrenin, the learned head of the Jesuit mission in the imperial court of Peking, started in 1729 with de Mairan's inquiry about the failure of the Chinese to make progress in the sciences.[18] Once more the fathoming of the question did not go beyond recording the differences between the curious European mind and the indolence of Chinese mentality in matters abstract and speculative.

The explanation was as superficial as it was fraught with inconsistency. Clearly, the curiosity of Europeans could not be looked upon as a feature impervious to further inquiry. Moreover, Father Parrenin was very much impressed by Chinese civility and moral philosophy which he traced back to ancient Chinese monotheism.[19] In this he merely voiced a pattern already stereotyped in the Jesuit China reports of which new, monumental, and scholarly instances saw print during the years when the de Mairan-Parrenin correspondence took place. The most important of them was Father Jean-Baptiste du Halde's *Description géographique, historique, chronologique, politique et physique de l'Empire de la Chine et de* la Tartarie chinoise (1735) in four folio volumes.[20] It gave powerful support to the pattern in question which had been accepted far beyond Jesuit circles by such disparate thinkers as Leibniz and Voltaire. Leibniz, in fact, urged, at the time when Father Lecomte's *Nouveaux Mémoires* was published, that China send missionaries to Christian Europe to teach the art of thinking straight in matters of ethical philosophy and of natural theology.[21] Half a century later, Voltaire heaped praises on the Chinese on the same count in his *Esprit des mœurs*.[22] Voltaire would not, of course, have agreed with Leibniz on continuing Christian missions in China. For Voltaire, the Chinese excellence in ethics and natural theology was a proof that the Gospels were not needed at all for upright living and thinking.

But Leibniz, Voltaire, and many others were at one in overlooking the inconsistency of their praise of the Chinese. Excellence in moral philosophy and natural theology certainly was an excellence in abstract thinking, the very capability which they found the Chinese to lack when it came to the sciences. The same capability stood also in stark contrast to the extreme superstitiousness of the Chinese which Voltaire was willing to recognize. He seemed, however, to sense the incongruity of his position because ultimately he blamed the cold climate of China for the backwardness of astronomy there. He found the origin of astronomy and of science in India, precisely because warm skies there, so Voltaire argued, permitted a sustained observation of the heavens.[23]

Such was a perfect case of armchair philosophy about distant lands and cultures but a philosophy facile enough to be digested in the climate produced by the *philosophes*. They were certainly uninterested in looking into the possible ramifications of what even Voltaire admitted, namely, the extreme superstitiousness of the Chinese. Spreading interest in the *philosophes* stifled interest in publications other than theirs. When de Mairan published in 1758 his correspondence with Father Parrenin on the question of science in China,[24] it produced no stir in the salons. There the *philosophes* celebrated themselves, confident of riding the wave of the future, a future based on reason and especially on its scientific kind. They all dreamed about a heavenly city to be set up on earth mainly by relying on science. The dream was largely a French dream even though its most memorable phrasing came from an Englishman, Joseph Priestley, who tellingly enough had to leave England. In speaking of the end of the world 'glorious and paradisaical beyond what our imagination can now conceive',[25] Priestley, as could be expected, simply brushed aside the question of a beginning. Once more the euphoria of being on the threshold of heaven made one forgetful of the path leading there, even though the path could logically invite questions about the long path of science itself. Of that long path only its allegedly final and perfect stretch was kept in mind, all the more so as it was located in France, the land of geometers and *philosophes*.

To an astonishing extent it was true that in less than two generations after the publication of the *Principia*, Newtonian science had been transferred from England to France. Condorcet was not exaggerating when in his article on integral calculus, which he contributed in 1776 to the supplement of the *Encyclopédie*, he proudly noted 'the restoration to the continent of Europe and especially to France of that superiority that Newton had bestowed upon England'.[26] In the same year Condorcet became perpetual secretary of the Académie, a position from which he spoke in a commanding tone. The once timid and disturbed teenager, who had exchanged faith and salvation in Christ for faith in mathematics and salvation in mankind, had recently been transformed by the conversational skill of Mlle L'Espinasse into the self-possessed prophet of a new age which he now was determined to usher in. The effort demanded a vision as well as an instrument. The latter was exact science, or more specifically the application of calculus to human affairs in the form of the calculus of probabilities which constituted Condorcet's main contribution to science. The vision, a mankind subjected to the dictates of science and enjoying its benefits, was set forth in Condorcet's swan-song, the *Esquisse d'un tableau historique des progrès de l'esprit humain*.[27] Composed in the closing months of 1793 and in the opening months of 1794, the *Esquisse* had in fact been in the making for twenty years. Its origin was a sketch on the effect of printing on scientific development, or a history of science in disguise. Its intended publication was to serve in 1772 Condorcet's plan to capture the post of perpetual secretary.[28] The ambitious sketch never was developed, though without any apparent loss for the historiography of science.

Although Condorcet took history for 'man's confession',[29] once he had turned into the self-righteous *philosophe* there remained in substance nothing to confess on the part of the living, that is, the *philosophes*. They could merely speak of their own virtues and of the vices of those many who had preceded them and had not had the good fortune of anticipating them. For all the barrenness of their view of history, the *philosophes* did not want to be separated from the historical past because such a separation would have meant a separation from that nature which they took, whatever their deistic lip-service to nature's Author, for the ultimate entity. A proof of this was their enthusiastic espousal of sensationism, a philosophy of mind and knowledge English in origin but appropriated by Condillac for the French, just as Newtonian science was successfully claimed by Clairaut and d'Alembert. The praises heaped by the Encyclopedists on Bacon's empiricism were so many votes for that sensationism. Adopted as a guideline in the historiography of science by Jean-Étienne Montucla, empiricism narrowed the horizon there to a recitation of scientific success, a situation that prevailed for almost two hundred years following the publication of his *Histoire des mathématiques* in 1758.[30] Its two large volumes also contained the history of mechanics, optics, and astronomy, which Montucla saw elevated through their heavy reliance on mathematics to the rank of exact science. The work had for its motto Bacon's dictum that as many pass by, science would progress (multi pertransibunt et augebitur scientia).[31] This was a subtle indication that in Montucla's treatment of the subject there was no natural room for questions about the complicated growth of science, let alone about its baffling stillbirths and one viable birth.[32] Owing to the great success and popularity of Montucla's work, which four decades later saw a second edition in four similarly large volumes,[33] oversight of such questions became a pattern in most histories of science until very recently. They were usually based on a more or less explicit empiricism, which proved itself to be an inadequate basis whenever those questions were given some attention.

This was already the case with two essays of Turgot written in 1749. One was a 'Philosophical Review of the Successive Advances of the Human Mind', the other, considerably longer, was entitled, 'On Universal History'.[34] The former was delivered to students at the Sorbonne. Its manuscript together with that of the second were already in private circulation before they came to the notice of Condorcet, who as a perpetual secretary of the Académie was an admirer and close collaborator of Turgot, especially when the latter was in charge of French finances. The two themes that set the tone of Turgot's essays are sensationism and the unlimited perfectibility of the human race. About the latter Condorcet was to give generous credit to Turgot in his *Éloge* of him.[35] The unlimited perfectibility of the human race was, of course, a tenet of sheer metaphysical faith, which showed that materialism is spiritualism in disguise and, like any Ersatz, is bound to reveal its inadequacy. Where this logic can best be seen in Turgot's sensationist and pseudo-metaphysical explanation of cultural and scientific history is in the ambivalent role which he gave to genius, scientific and otherwise.

Turgot would have been more consistent had he ignored the topic of genius completely, but he could not do so since he was one of those *philosophes* who looked upon themselves as geniuses. But on the basis of a sensationist philosophy the genius could be but the chance occurrence in a very large number of ordinary minds, all of which were produced by sensations. Or as Turgot put it: 'Genius is spread through the human race very much as gold is in a mine. The more you take, the more metal you get. The more men there are, the more great men you will have, or [at least] the more men capable of becoming great.'[36] The crudeness of Turgot's explanation of genius received a stunning illustration a few pages later where he took up the question of the rise of genius in painting. Once more the logic was that a widely shared mediocrity will issue in the excellence of a few. Dialectical materialists said nothing new a century later when they proclaimed the fallacious dogma that quantity generates quality. Fallacy was waiting in the wings as Turgot declared that 'in all crafts where bad workers cannot gain a livelihood, and second-rate workers are not comfortably off, great men are not created'.[37] In the same breath Turgot referred as an illustration to 'our [French] painters of Notre-Dame bridge [in Paris] who supply all the little village churches with pictures, and constitute an indispensable nursery for the rearing of a few great painters'. The worst of fallacy was, however, still to come. It made its appearance in that instinctive attachment to their Roman Catholic heritage of many French intellectuals who had parted with faith in Christ in which they were brought up. 'I do not believe', Turgot declared, 'that one can name a single painter of history who is at all well known who was not a Catholic'.[38]

This was a declaration that could not be saved by further explanation. For there was no saving grace in Turgot's dismissing Dutch painting as limited to landscapes, seascapes, and rustic scenes, as if Rembrandt, Vermeer, and others had not existed. It was, of course, true that, as Turgot remarked, 'the English for many years have spared nothing to acquire beautiful pictures'. There may even have been some truth in Turgot's claim that until then the English 'had not been able to produce a single great painter'. According to Turgot, 'the English pay only for good pictures' but do not paint them, because they do not paint mediocre and bad pictures. Turgot traced the absence of mediocre and bad painters in England to the banning by Protestants of images from churches. The English thereby deprived themselves, Turgot noted, 'of the means of supporting bad painters and even second-rate ones'.[39] What was particularly wrong not so much with English painting but with Turgot's reasoning was that within a generation or so there appeared in England a galaxy of painters whom the Continent quickly recognized as being first-rate. It is still to be shown that Gainsborough, Turner, Blake, and others were the fruits of widespread mediocrity.

One can only speculate about the chain of inconsistencies had Turgot tried to document the existence in the misty Baltic lands of many mediocre Copernicuses that preceded the great one, whom, tellingly enough, Turgot failed to mention when listing the great initiators of science in modern times. His list started with Francis Bacon, never a scientist. Nor did Turgot face up to the problem which his genesis of a few giants from many dwarfs presented in connection with China, a country with an impressively vast population even in his times. If sensationism was true, then China had to be the leader not only in the number of sensations experienced by men but also in that science which those sensations were believed to generate in a fairly automatic fashion. China was anything but a leader in science. In Turgot's graphic description China resembled 'those trees whose trunk has been topped and whose branches grow close to the ground; they never escape from mediocrity'.[40] This fate befell the Chinese, because they had become too proud and respectful of their rapid initial progress. Turgot felt that slow progress would not have invited a similar pitfall.

But more fundamental than the pride invited by rapid progress was that rapid progress itself which science achieves through geniuses, a point that leads to the relation of brain to genius as imagined by Turgot. He started with the slight differences in the 'fortunate arrangement of the fibres of the brain, a greater or lesser degree of power or refinement in the organs of sense and memory, a certain degree of quickness in the blood'.[41] It was among these differences that education could chance upon the potentially most promising one and give it the needed impetus. Education was, however, in part at least, an exercise in ideas which sensationism could hardly account for. Again, the disparity could appear but enormous between the education Newton received in Cambridge and his scientific creativity, which asserted itself sometime before his formal education was over. It was precisely on the basis of sensationism that the rarity of genius was difficult to account for, and to accept Turgot's claim that 'had Newton died at sixteen we would still perhaps be ignorant of the true system of the world'.[42]

That system of the world was the fruit of an extraordinary reach beyond what is sensed, a reach that demands a mind worthy of being called such. Whatever their commitment to sensationism, Turgot and his fellow *philosophes* spoke profusely of the mind as they hailed the century of reason, their own century. But they hailed reason with too much passion to let its light truly break through. Turgot's portrayal of the progress of enlightenment was in fact an apotheosis not of reason but of passions. After all, it was impossible to gainsay the existence of scientific progress long before the onset of the age of Enlightenment. The credit for that progress could not therefore go to reason. As Turgot portrayed it, the passion was that of the tumultuous movement of each part of a huge army which makes its way through the obstacles without suspecting a grand design conceived by a military genius. Such was a scenario based on non-sequiturs certainly glaring in a framework in which 'enlightened' reason was the sole explanation: 'Thus the passions have led to the multiplication of ideas, the extension of knowledge, and the perfection of the mind, in the absence of that reason whose day had not yet come and which would have been less powerful if its reign had arrived earlier'.[43]

The genius or leader mentioned in the context was personified Philosophy, one more instance of the self-styled rationalist's relapse into animism and mythology. Philosophy and Reason written with capital P and R in Turgot's time were replaced a century later by such passions as Instinct and Struggle, also duly capitalized as befitted the personified abstractions of a mind forgetful of its privileged status in nature. And even if the leader mentioned by Turgot had been something more than a personified Philosophy, he was certainly not the Christian God, the sole God capable of a providential leadership of all. The only beneficial role which Turgot allowed for Christianity and Gospel in cultural history was the taming of some passions and the nurturing of brotherly love, but he was quick to add that first Christianity but afterwards philosophy 'taught man to love all men'.[44]

Whatever the effectiveness of philosophy in replacing the Gospel as a source of brotherly love, philosophy as conceived by Turgot proved to be most ineffective and unconvincing to cope with scientific and cultural history. His essays were unwitting proof that any age which looks upon itself as the epitome of this or that perfection inevitably loses its sense of history, which is always a march and never an arrival. Turgot and the *philosophes* based their belief that progress had arrived at its ultimate phase on the presumed perfection of physical science. Turgot's analysis of progress amounted to the claim that the true form of physics had already been ascertained to such a degree 'of certainty that today it is only the details which remain in doubt'.[45] Condorcet said much the same in his *Sketch*,[46] enough to undermine its claim of being an enlightening account of cultural and scientific progress. His enterprise was doomed to failure through the very starting phrase of the *Sketch*, a flat endorsement of sensationism. On its basis, Condorcet reached by the end of his outline of the first phase of human development the revealing conclusion which could hide its self-defeating character only to those ready to be duped by the *philosophes*.

The conclusion stated that until the age of *philosophes* came, mankind had remained subject to the deplorable 'separation of the human race into two parts; the one destined to teach, the other made to believe; . . . the one wishing to place itself above reason, the other humbly renouncing its own reason'.[47] With such sweeping categorization of much of intellectual history Condorcet's essay could not even be a sketch of it but only its sketchy abuse aimed at disclosing 'the cause of the credulity of the original dupes, and the cause of the crude cunning of the original impostors'.[48] Those who two hundred years after Condorcet still write intellectual history with his blinders on their eyes, do not fail, of course, to reassert about the present what he had said of his own days. According to Condorcet the foregoing separation that had already been imposed amongst the least civilized savages 'by their charlatans and sorcerers' was still being advocated by 'priests in the closing years of the eighteenth century'.[49]

Condorcet's unremitting and virulent antagonism toward Gospel in particular and religion in general is of interest here only inasmuch as it deprived him of seeing the deepest source and ultimate perspective of what was the very core of his not-so-scientific vision of a secular religion. That core was his belief in the fully rational interrelation of all in a single whole,[50] and for a single end. This in turn permitted or imposed, the 'referring of all to a single point of reference'.[51] Condorcet was blind to the fact that such an outlook, which was the basis of science itself, was a metaphysical one, in fact a religious Weltanschauung. But he was not entirely unaware of the fact that, strangely enough, this very same outlook had from the earliest days of mankind been advocated by those 'charlatans and sorcerers'[52] and reached a very purified form in the hands of scholastics strongly tied to the priestly caste. Clearly the scholastics, or medievals, did more than perfect the art of logic if Condorcet could acknowledge his and the *philosophes'* debt to them on tenets such as 'the more precise notions concerning the ideas that can be determined about the supreme Being and his attributes; the distinction between the first cause and the universe which it is supposed to govern; the distinction between spirit and matter; the different meanings that can be given to the word *liberty*; what is meant by *creation*; the manner of distinguishing the various operations of the human mind, and the correct way of classifying such ideas as it can form of real objects and their properties'.[53]

All these points indicated that the scholastics were interested not merely in notions about reality but about reality itself and in a way which was indispensable for a scientific study both of separate things and of their totality, the universe. For Condorcet the foregoing scholastic method, which led to specification of those precious affirmations about God, man, and the universe, could only retard the progress of natural science. Such was a baffling inconsistency, of which many curious smaller examples can be found in the *Sketch*. Thus, for Condorcet, Bacon was both the genius of scientific method and also one who sadly lacked genius for doing science.[54] The root of that inconsistency was Condorcet's addiction to sensationism, a straitjacket for his interpretation of scientific history. Only a sensationist could be so senseless as to state that Copernicus had eradicated from heliocentrism 'anything that was repellent to sensory experience'.[55] It wholly escaped Condorcet that according to the historical record it was precisely Copernicus' failure to do so that made the acceptance of his *On the Revolutions of the Heavenly Spheres* a very slow and very gradual revolution indeed. But for Condorcet the progress of science was 'swift and startling'[56] from precisely the times of Copernicus on. Its opposite, namely, the slow progress of science or simply its standstill in other places and times, was just as unexplainable within a sensationism unable to do justice to mind, let alone to genius. The most monumental of those standstills was exemplified in China. In Condorcet's *Sketch* the frustrating performance of the Chinese of old constituted the most telling case of 'the lengths to which these institutions [of priestly castes and impostors] can carry their destructive powers over human faculties'. But even when Condorcet's sensationism is considered apart from his antireligious fanaticism, 'the eternal mediocrity' to which Condorcet condemned the Chinese was just as little explained as was their erstwhile feat to have in the beginning 'outstripped all the other nations in the arts and the sciences'.[57]

Herein lay a problem fully perceived and aired in print almost two decades prior to Condorcet's writing the *Sketch*. In fact he was fully aware of Bailly's *Lettres sur l'origine des sciences* published two hundred years ago, a work presented by Bailly as a series of letters addressed to none other than Voltaire.[58] The chief motivation behind the *Lettres* was Bailly's resolve to clear up the puzzle presented by the presumed origination of science by the Chinese or other ancient nations, and by their inability to keep advancing that very science which they had originated. The problem was concisely put in the question raised by Bailly already in his *Histoire de l'astronomie ancienne*: 'Though there may be nations so unfit to march as to be part of the advance of science, would the nation that has once become part of it by the momentum which it imposed on itself ever lose that momentum and come to a halt?'[59] The question invited a resounding no and this in turn prompted Bailly to speculations that made his mind journey from science to Mongolia.

In that intellectual odyssey so uncharacteristic of a resolute rationalist he was guided by two erroneous presuppositions, both of which had to do with the narrow views which the Enlightenment imposed about history and not in the least about scientific history. In their illusion about the perfection of Newtonian science and especially about its explanation of planetary motions (in which they saw the system of the world), the men of Enlightenment, like Bailly, himself an astronomer, overlooked its painfully long genesis from Copernicus to Newton. Contrary to their belief that genesis was not an automatic process. They mostly saw Newtonian science as something ready-made, almost born in the form of an adult. Consequently, in their survey of ancient science they looked for its complete anticipations, however embryonic. The ancients offered, however, only what Bailly aptly called debris of science.[60] Moreover, he found such debris, mostly astronomical (calculation of the length of the year; construction of calendars; specification of lunar and planetary cycles and the like), remarkably similar. Was it reasonable to assume that all ancient cultures failed in the same way, or that they had taught one another about those fragmentary items? Was it not more reasonable to assume, Bailly asked, that all that debris scattered across ancient cultures were the remnants of a truly scientific system devised by a nation no longer in existence?

In going beyond all presently known ancient cultures, Bailly had no wide choice as to the direction to take. The widespread belief that Chinese culture was the oldest in existence forced him to look north of China, toward the largely deserted steppes somewhere, as he put it with an affectation of astronomical precision, between the 50° and 60° parallels.[61] It was there that he located his mythical, ingenious people, the originators of a genuine science, who for some unknown reason were uprooted and scattered to the south. A combination of dislocated masters and less ingenious pupils, such as the Chinese and the Hindus, resulted, so Bailly argued, in a deterioration or debris of science, transmitted later to the Chaldeans and Egyptians.

In all likelihood, Bailly would have pushed the flourishing of his mythical nation much further back than the fourth millennium B.C. But he had to make concessions to the theologians of the Sorbonne on biblical chronology and especially on the dating of the deluge. But in crediting his mythical nation with so much intellectual and moral perfection Bailly knew that he was rebuffing the theologians on a far more fundamental issue, namely, the question of original sin. It was that notion which was mostly resented by the men of Enlightenment determined to spread the rationalist gospel that man needed no redeemer except himself. Between this gospel and the Gospel there could be no meeting of minds.[62] Such is the real root of the irreconcilable dislike of the *philosophes* for the Middle Ages, an age all too conspicuously rooted in the Gospels. It was that dislike that prevented them from perceiving the medieval beginnings of the only viable birth of science. The same dislike also prevented them from seeing Greek science for what it truly was: a stillborn science, and from seeing in the debris of science in other ancient cultures so many evidences of other stillbirths of science.

In their failure to see the only viable birth of science and its several stillbirths, the *philosophes* were forced to look at intellectual history as being in need of a gigantic revolution which they were most eager to bring about. It was they who made the expression 'révolution scientifique' or 'révolutions dans les sciences' a stylistic cliché[63] which like other clichés are more misleading than enlightening. Today the same cliché still confuses discourse about the origin of science as much as it did two centuries ago when it became popularized by those ushering in the French Revolution. Bailly was a chief among them and also one of those *philosophe*-scientists who lost his head to the Revolution. In a sense he had already lost it when writing a book, the *Lettres*, on the origin of science and credited that origin to a mythical nation about which he knew, as d'Alembert remarked, everything except its existence.[64] Such was the price demanded by the inner logic that led even a talented historian of astronomy, like Bailly, from the historic reality of science to an imaginary nation in Mongolia. Bailly's equally imaginary journey was all the more futile because the documents of the origin of science were in his own European backyard. Being highly despised medieval manuscripts they all were gathering dust in the soon-to-be-dissolved monasteries. To make the irony complete the most significant of those manuscripts had been written in Paris and still were located there. Unknown to the *philosophes*, Paris was a city which had diffused truly creative light long before it became a City of Light with a beacon which pointed to Mongolia as the cradle of science.

THE RUBBISH OF HISTORY

A programmatic preoccupation with science during the seventeenth century was largely an English and during the eighteenth century mainly a French phenomenon. Attention could therefore be confined to England and to France respectively in the first two lectures without doing much violence to the historical record. The nineteenth century allows no such shortcut. Of the three powers, England, Germany, and Russia, that opposed Napoleon at Waterloo, England and Germany were to play a leading role in nineteenth-century science, posing a challenge to the leadership which France tried to retain as her eighteenth-century heritage. Not surprisingly, the most significant reflections that were offered during the nineteenth century on the origin of science had three distinct origins, French, German, and English.

This order, French, German, and English, is not chronological, the kind of order to be given first choice in a historical essay. Both Spencer and Whewell were preceded by Hegel but not by Marx and Mach. As to the relation of Comte and Hegel, the chronological order gives priority to the latter, but Comte, who started publishing his *Cours* only after Hegel's death, represents a more direct tie with the eighteenth century. Of course, Hegel cannot be understood without Kant, but the *Opus postumum*, Kant's lengthy speculation on the true form of science, could not have been known to Hegel.[1] He unfolded independently that shocking set of precepts which followed from Kant's subjective idealism for the interpretation of science. To be sure, Kant did not turn his attention to the history of intellect (*Geist*), including the history of scientific thought, to the extent to which Hegel did. But Hegel came of age only in Marx and Engels, both of whom followed Comte in at least a chronological sense. As to Spencer, whose ideas will lead to the concluding part of this lecture, his positivism owes much to Comte.

The chief reason for taking Comte as the starting point of the nineteenth-century part of our story lies in Comte's strong ties to Turgot and Condorcet, protagonists of that story during the eighteenth century. As to Turgot, the first to portray scientific history and the question of its origin with an eye on bringing about a social Utopia, Comte's ties with him lead through Saint-Simon whose collaborator young Comte was for a few years. Those ties were in a sense severed when Comte published in 1822 his *Plans des travaux scientifiques nécessaires pour réorganiser la société*, better known under the title of its second edition (1825), *Système de politique positive*,[2] a smallish book which contains all the basic ideas he was to articulate later in massive volumes during the remaining thirty years of his life.

When in 1824 Comte wrote of the time he spent with Saint-Simon as 'lost years',[3] he did so in the realization that Saint-Simon could only be a tie to that Turgot whom Comte too came to see as 'machine Turgot',[4] so nicknamed for his fondness for mechanical contraptions. The label certainly fitted the fact, for most of his active life Turgot was not so much thinking about society as tinkering with it. Comte, however, aspired for a philosophical grasp of what an ultimate social reform presupposed and this is why he discovered Condorcet. Tellingly enough, Comte, the one-time *répétiteur* in calculus at the Polytechnique, did not look for inspiration in that Condorcet who, for instance, wanted to base the proper functioning of courts on considerations relating to probability calculus.[5] For Comte, society was a living entity and as such, like all living beings, impervious to exact mathematical analysis. This, however, did not mean for Comte that living entities, including society as living through its history, were not governed by scientific laws. In addition to the laws encountered in the exact sciences he advocated a broader sense of scientific laws, the sense embodied in his law of three states, or the law of inevitable advance from the religious to the metaphysical, and from there to the scientific or positivist world view. That broader notion of scientific law was clearly present in Condorcet's *Sketch* and was in fact far less scientific than metaphysical and religious. No wonder that Comte used the expression 'spiritual father' in speaking of Condorcet.[6] The expression tells not only on Comte, the founder of the positivist church, but also on Condorcet, author of an 'Almanach antisuperstitieux'. It anticipated in a negative sense[7] Comte's calendar of the saints of positivism in his *Catéchisme positiviste*.[8] Condorcet was hardly displeased with the enthronement of the Supreme Being in Notre-Dame, an event which Comte could hardly help recalling as he predicted in 1851 that not later than 1860 he would in the same Notre-Dame 'preach positivism as the only real and complete religion'.[9]

These details are recalled about Condorcet and Comte not only to put in proper light their so-called rationalism but also to help understand why Comte had failed completely with his ideas on the history of science and on its origin. He failed, because like Condorcet he advocated rationalism as a gospel destined to replace once and for all *the* Gospel. It mattered not that, unlike Condorcet, Comte wanted to retain as much as possible from the cultural structure grown around the Gospel. The 'Catholicism minus Christianity', to recall

Huxley's concise summary of Comte's programme, offered an outlook on history not essentially different from that of the *Sketch* of Condorcet.[10] While Huxley's phrase referred only to the contents of the *Catéchisme positiviste*, the fifth volume of the *Cours de philosophie positive* written more than ten years earlier would have already justified the same appraisal. In the *Catéchisme* the first of the thirteen months of the positivist year was the month of theocracy dedicated to Moses with Numa, Buddha, Confucius, and Mahomet being celebrated on its four Sundays (Humanidi). Egyptian, Chinese, Tibetan, and Jewish prophets filled the weekdays, but Jesus Christ was not mentioned, a procedure all the more revealing as John the Baptist was not forgotten by Comte.[11] A dozen or so years earlier Comte had already given the same silent treatment to Jesus of Nazareth when in the fifth volume of the *Cours* he spoke of 'the real or ideal founder' of Christianity without mentioning him by name.[12] This crude historiographic pattern was initiated by Voltaire, who in his universal history jotted down the name of Jesus only once and only after he had reported Constantine's crossing of the Milvian Bridge.[13] Such was the rationalist historian's flight from historical reality to the land of fiction, as distant from history as from rationality.

Separating Catholicism from Christ and reducing Christ to a practical nonentity was part of Comte's strategy to present the ten long medieval histories from Augustin to Dante as a phase of history imbued with an ethical concern which was not so much the message of Jesus of Nazareth but of Paul of Tarsus. Comte needed that moralizing phase if his law of three stages was to retain some semblance of history.[14] No historian of any responsibility would today recommend the lengthy section on the medieval period or any other part of the fifth volume of the *Cours* for what it was meant to be by its author, a reasoned survey of cultural history. For him history existed only as an illustration of the positivist dogma about the possibility of determining the exact laws of social life. He believed that the possibility of those exact and immutable laws was proven, once he had established the existence of exact, that is, immutable laws in the physical world. Such was the Comtean progress from ordinary physics to social physics, which in fact meant an enormous regress in the cultivation of the history of science, to say nothing of physics.

It has been customary to deplore Guizot's decision not to accept Comte's proposal about a chair to be established in the Collège de France for the history of science, the first chair of its kind and with Comte as its first occupant.[15] In 1832 Guizot might still have appeared a narrow-minded (and reactionary, as some would say today) administrator, but the publication of the first three volumes of the *Cours* within six years made it clear that Comte would not and could not write history of science. Any keen observer might have reached this conclusion already by 1822, when Comte published his *Prospectus*, and certainly after Comte's oral delivery in 1828 of the first two lectures of the *Cours*. They revealed a systematic, dogmatic mind which could only straitjacket the undogmatic though not illogical course of intellectual history.[16] Straitjacketing it certainly was for the history of science in more than one sense. First, what came to be confined within the straitjacket was first painfully constrained and then denied its future development. The historical life of the various branches of science certainly did not justify the strained proportion about them as given in those three volumes. The first was almost entirely devoted to reflections on the mathematical sciences with heavy emphasis on infinitesimal calculus, the subject which Comte taught at the Polytechnique. The second volume was taken up by an astronomy largely reduced to celestial dynamics and by a division of physics into five parts according to the five senses of man. All the rest of the sciences, chemistry, biology, physiology, and psychology, were crammed into the third volume.

Such was a straitjacketing demanded by the Comtean idea of exact or immutable laws whose existence could not be argued with any prospect of success outside astronomy conceived as celestial dynamics. Demonstration of that immutability was the foundation of Comte's subsequent discourse on the imminent onset of positivist society as the final and immutable stage of social development.[17] Clearly, a Christ who placed heaven beyond earth revealed a far better historical instinct than Comte and several other prophets of the nineteenth century who conjured up a heaven on earth. While there was nothing in the Gospel of Christ that would deny progress to science, Comte could but deny to astronomy the exploration of the stellar realm, because he felt that new discoveries about distant realms might pose a threat to the reputedly non-revisable character of already established laws. He restricted cosmology to the study of the solar system,[18] although even there he did not approve going beyond the five visible planets. The enthusiasm that greeted the recent discovery of Neptune was for him insane[19] and he slighted Herschel's findings of binary stars.[20] What he offered on electricity, chemistry, biology, and psychology is a storehouse of statements befitting those to be put in a straitjacket.[21] For Comte, Gall, the founder of phrenology, was a scientist comparable with Copernicus, Galileo, and Newton.[22]

Denying a future life or development for science meant a denial of its having originated in a manner resembling a truly living entity. Having inherited the dogma of the Enlightenment about the invincible ability of man to get rid of all errors, Comte could not help being mesmerized by the breaking through of light in full which he identified, in accord with an already hallowed cliché, with Bacon, Galileo, and Descartes.[23] In his rationalist perspective there was only light versus darkness, a perspective certainly dogmatic though not historical. Tellingly enough, the fifth volume of the *Cours*, a historical illustration of the law of three states for social physics, was preceded by a fourth devoted entirely to the dogmatic outline of the same law in its relation to society. With such overtly *a priori* or rather dogmatic approach to history everything therein could be safely ignored which did not seem to be connected with the breaking through in full of the light of reason. Applied to the history of science this meant that with the exception of the Greeks all other ancient cultures could safely be ignored. Intellectual history, Comte declared, ought to be concerned with the intellectual élite among the various races,[24] an easily racist standpoint which Comte might have spared himself had he been willing to consider the Gospel truth about a truly élitist dignity universally shared by all members of the human race as distinct from all other races, a distinction which Comte was eager to abolish.

His resolute dismissal of the Gospel and his explicit dismissal of ancient China and India as totally irrelevant for the history of European culture[25] and science are not therefore as disconnected as they may appear to be. Being a gospel about a social Utopia just around the corner, Comte's doctrine denied a future to history, intellectual and social, and denied to itself by the same stroke the perspective into the past which, if it is a genuine past, includes also its logical basis, namely, its genuine origin. Such is the chief instructiveness today of reading Comte, who until very recently could pass for an instructive historian of science.[26] In reading him one is treated only to positivist dogma and to nothing really positive about history, let alone the history of science. A revealing confirmation of this is offered by the voluminousness of Comte's writings. They are conspicuously barren of references either to original sources or to secondary literature. Being the kind of dogmatist he was, Comte did not need historical sources to learn from them. It has been an open secret during Comte's life that he had stopped reading old as well as new books years before the sixth volume of the *Cours*[27] was published in 1842.

Any writer who stops reading may still continue as a poet or novelist but he certainly undermines his claim to be considered a historian. Unfortunately, Comte was not the only chief nineteenth-century interpreter of cultural history who either stopped reading or failed to read what he should have read while speaking authoritatively about the history of science in its broadest perspective. Hegel was one of the latter. Specialists of various aspects of Hegel's thought could easily give a wealth of documentation similar to the one that constitutes Michael Petry's three-volume commentary on Hegel's interpretation of the natural sciences.[28] But even a modest acquaintance with the history of science can make it clear that Petry's industry in establishing a connection between Hegel's sundry dicta and his scientific sources proves beyond any doubt that though Hegel read voraciously, he did not read the right sources. On hearing such an indictment Hegel might shrug it off with a reference to his distinction between the temporal and the dialectical development,[29] a distinction which would safeguard the correctness of the laws of his dialectic even if his reading of the temporal, the contingent, the material, and the scientific were to be proven gravely at fault.

In the explanation and justification of those laws Hegel's views on the Gospel play a crucial part. According to his deepest conviction it was in the Gospel, in the teaching of Christ, that the conflict between finite and infinite was truly transcended. The teaching in question centred on love, but —and here enters the Hegelian dialectic in its most telling form—Christ quickly realized that He could persuade about the meaning of love His own people—a people convinced of having been taught by God—only by presenting his teaching about love as coming directly from God. This is why Christ, with a cunning worthy of all true adepts of Hegelian dialectic, right and left, presented His own authority as being equal to that of God though He surely did not consider Himself as being equal to God. The divinity of Christ was, according to Hegel, an illusion produced among His disciples, an illusion which had befogged eighteen centuries.[30] But if historical reality could so brazenly differ from the logic of Hegelian dialectics, it could have but secondary significance for Hegel and Hegelians.

Once reduced to that secondary level, history could be written in bold strokes, which Hegel certainly did in his *Philosophy of History*. There ancient history dominates the presentation possibly because it is far easier to theorize about the little known facts of the distant past than about the well-known facts of recent centuries. Thus Hegel spoke at length about the ancient Chinese, Hindus, Egyptians, Persians, Greeks, and Romans, while he was astonishingly short on the moderns.[31] To be sure, his work is a *philosophy* of history, yet its historical basis is much too weak to let it pass for a philosophy of *history*. It closely parallels his discourses about the concrete, contingent, and material. His many dicta about the cosmos were such as to justify the writing by McTaggart of a classic monograph on Hegel's cosmology in which there is not a single detail about the real cosmos which science deals with. It was the same McTaggart, not a believer himself, who astutely remarked that as a presumed ally of Christianity Hegelianism is 'an enemy in disguise—the least evident but the most dangerous. The doctrines which have been protected from external refutation are found to be transforming themselves till they are on the point of melting away'.[32] Here again Gospel and science are once more in the same boat. The Gospel with the doctrine of the Word made flesh at its centre makes no sense once man's sense of flesh and blood reality has been diluted by the solvent of Hegelian dialectic. Nor can science retain its sense of reality on the basis of Hegelianism. This is why as a presumed ally for science or for the interpretation of its history Hegelianism is 'an enemy in disguise, the least evident, but the most dangerous'. This point is still to be emphasized, although its truth was already in evidence in the ferocious attacks of *Naturphilosophes* on scientists interested in a given nature, too hard in its givenness to melt away under the fervour of their inquiry.

Clearly a nature which is the dialectical product of the unfolding of the spirit can have its corresponding explanation only from a strictly introspective mind. This is why in Hegel's view true science was still to be written and this is why for Hegel and for Hegelians no serious question can arise about the origin of science. Hegel's cavalier handling of Copernicus, Galileo, and Newton amply proves this point. The science, or physics to be specific, still to be written was, in Hegel's words, a 'higher physics', a physics far transcending the merits and perspectives of what Newton had produced. Hegel's first publication already had served an example of what that physics looked like. The publication, Hegel's *Habilitationschrift*, came to a conclusion by setting an *a priori* limit to the number of planets around the sun. The number in question was based on what Hegel called the Pythagorean sequence of numbers and it could but make appear foolish the already avid search for a planet between Mars and Jupiter.[33] The *Habilitationschrift* was just leaving the press when word about the sighting of the first asteroid began to electrify Europe. Prince Ernest of Saxa-Gotha could hardly suspect the historic validity of his words when he characterized to F. X. von Zach, a leading German astronomer of the day, Hegel's booklet as 'monumentum insaniae saeculi decimi noni'.[34] The booklet was a monument that cast indeed a long and dark shadow into the future in many ways and certainly for science whenever science found itself in an ambience where adherence to Hegelian dialectic has become mandatory.

The first stretch of that long shadow was *Naturphilosophie* as it developed into a crusade against scientists during Hegel's undisputed domination of the German academic scene in the 1830s.[35] The second stretch came with the rise of the Hegelian left. Its protagonists' first quarrel was with the Hegelian right, and Engels must have had Hegel, the Hegelians, and the neo-Kantians in mind as he complained in a letter in 1894 that 'it has become a custom in Germany to write the history of sciences as if they had fallen from the skies'.[36] The unfolding of the absolute in Hegel's or in the Hegelians' mind certainly gives the impression of coming straight from the upper regions, in a curious contrast to the exceedingly human manner in which biblical revelation was given to man. The narrative of the Gospel was indeed so far removed from the rarefied higher atmosphere where the Hegelian spirit loves to roam, as to provide endless pretexts for anyone contending that Christ had no heavenly origin.

The most virulent among these contenders come from the

ranks of the Hegelian left, opponents of any higher level, be it purely intellectual. Their hostility to the Gospel is a piece of their aversion to that creative intellect which is the life-blood of science. Their prolific discourses about science and its origin are as revealing as the instructions given on that score by Engels to Marx who invariably turned to the former for advice in matters scientific. It was a procedure worthy of that dialectic in which the negative can provide the positive. Those familiar with Engels' crude pontificating about science in his *Dialectics of Nature*, with his calling Newton an inductive ass, with his invectives heaped on the formulators of the second law of thermodynamics, and with his continual recourses to the 'good old Hegel' as the most insightful man of science,[37] will easily grant that the help Engels gave to Marx was not better than the direction which one blind man gives to another. To grant this even a layman moderately informed in the history of science will not find it necessary to read the *Dialectics of Nature*. It will be enough for him to read from the foregoing letter of Engels a few more phrases in which Engels placed the origin of science in practical needs and declared: 'we have only known any-thing reasonable about electricity since its technical applic-ability was discovered'.[38] An astonishing phrase worthy of a Marxist's blind faith in the truth of dialectical materialism. Clearly, it takes a blindly dogmatic mind to imply that there was nothing reasonable in the electrical researches of Frank-lin, Coulomb, Oersted, Ampère, Biot, and Faraday, all of whom worked before Kelvin designed an efficient trans-atlantic cable, Siemens constructed a workable dynamo, and Edison invented the electric bulb.

In that blind dogmatist approach to history even its most relevant data can become sheer rubbish, though Marxists have no liberty to dethrone in such a way that history which is to provide them with nothing less than a throne sometime before the advent of a classless society. In fact, it was not a Marxist who said 'I must candidly confess that I hate the rubbish of history'. Yet, since Ernst Mach, who uttered these words, still has the renown of a prominent historian of science, the context of his statement cannot be ignored lest the suspicion arise that he was quoted out of context. The context is a public lecture which he gave early in his career, a few years before he published his first essay on the history of science. Yet, the lecture given in Graz in March 1867, on the question 'Why Has Man Two Eyes?' contains in a nutshell the reasons why all Mach's subsequent writings on the history of science evidenced a narrow one-way vision in which, if not science itself, at least its history appeared hardly more than mere rubbish. The very first thing Mach did in that lecture was to tell his audience that the title as such was inadmissible within, as he put it, the 'Church of science'. Within the orthodoxy of science, Mach continued, one could only ask: 'Since man has two eyes what more can he see with two than with one?'[39] Mach was not yet through with one-third of his lecture when he gave the answer, namely, that because man has two eyes, he can see distance in addition to forms.

True enough, but then why did Mach need the remaining two-thirds of his lecture? Well, he needed them for his interpretation of cultural history which could but debase history to the level of mere rubbish. On that level the history of physics, Mach's vaunted speciality, had its origin, to quote Mach's words, 'in the witch's kitchen'.[40] This hardly did justice to the origin of physics, or of science for that matter, but was a logical consequence of Mach's sensationism, a kettle befitting that kitchen. The kettle of sensationism was useful for concocting sensory data into a subjective brew, but was unfit for extracting their intelligible objective content which raises man's sensations to a level incomparably higher than the one where animals carry on with their sensory acts. No wonder that the various steps of the development of the art of perspective, that is, the art of illustrating man's vision of depth, were for Mach not indices of an ability unique with man. Man's vision of depth was for him merely a case of the many various forms of vision to be found in the animal kingdom from insects through birds to monkeys.

Rarely was an ability of man handled with less view in depth and Mach had to ward off the spectre of superficiality with a facetious twist, part of his description of the universe as it would appear to a creature knee-high to a leaf of moss. The twist was Mach's warning, 'with thy puny bulk thou shouldst not joke with capillarity', addressed to that puny creature trying to explore a drop of water sparkling on that leaf. Phrases like this fit novels known as science fiction, the kind of novel that would have appealed to Mach. Had he been a novelist, he noted, he would not have chosen for his heroes any of his contemporaries, nor anyone from ancient times and places such as Pharaonic Egypt. He would not choose them, past or present, because as he put it, 'I hate the rubbish of history'.[41] He could but hate history because for his sensationism it was as indigestible as any rubbish would have been. Interesting though history may be as a mere phenomenon, he was unable to like it because, 'we cannot simply observe it but must also *feel* it'.[42]

This was certainly true. The understanding of history demands more than to notice facts. A historian must *feel* the facts, that is, he must be able to identify with them and more so with some facts than with others, or else his account of history will be a mere enumeration of facts, a process depriving all of them of that valuation which makes the historian capable of seeing *history* in mere events. As Mach unerringly sensed, this *feel* for history was tantamount to ascribing to man a special dignity, which he flatly refused to do and he based his refusal on science: 'In making man disappear in the All, in annihilating him, so to speak, they [the physical sciences] force him to take an unprejudiced position without himself, and to form his judgments by a different standard from that of the petty human'.[43]

Man certainly owes much to that science which is so apt to reveal to him his pettiness. But that very same science is also extremely effective in revealing something of man's unique grandeur which in no way could be gauged by Mach's sensationism. Moreover, the history of science also shows that man had to achieve conviction of his uniqueness before a viable science could rise as a unique phenomenon of history. Mach failed to perceive this connection precisely because of his hatred of Gospel and Christianity,[44] two factors which more than any others provided man with that conviction. This is a principal reason why Mach could not have become a historian of science even if he had wanted to. There was a truth much deeper than he suspected in his implicit insistence that he should not be considered a historian.[45] That he was not one was clear in that subtly devastating conclusion of Duhem about Mach's *Science of Mechanics*, namely, that its author sees the history of the subject in terms of its final phase.[46] That final phase was for Mach a phase wholly divested of any metaphysical, let alone theological, connotation. Consequently, he prevented himself from taking proper account of the growing evidence about a pre-Galilean science of mechanics and dynamics.[47]

Mach was too keen an intellect not to perceive that any acknowledgement on his part of the medieval roots of the science of motion would pose a serious threat both to his derivation of science from tool-making and to his hostility to metaphysics, theism, and the Gospel. He knew perfectly well what the real alternatives were, as can be seen in the same lecture on 'Why Man Has Two Eyes'. The lecture is anticipation in a nutshell of the message which Mach was to preach in all his career, much of which still lay ahead of him. In that lecture, in which he had poked fun at belief in Adam's fall, he ended with a warning put by Goethe in the mouth of Mephistopheles, hardly the proper teacher to instruct man to avoid, in the spirit of humility, thinking of himself as an individual whole rather than an insignificant part of the Whole.

Humility is the art of thinking in just proportion, but it was precisely that art that could not be had on the basis of sensationism which Mach embellished with the notion that thinking was the economical correlation of sensations. Mach did not, of course, wish to suggest that the correlation in question was the fruit of the mind's laborious work. Not admitting the ontological reality of the mind Mach could not credit the mind with anything and certainly not with that modest effectiveness by which it gropes its way amidst the welter of sensations towards the light of understanding. Indeed, the mind could at best be a mere passivity if Mach's claim was true that if all sensations were present to the mind the pursuit of science would no longer be necessary.[48] Happily for science, no scientist of any distinction paid attention to that claim of his. They obviously realized that if sensations constituted science, no scientist could ever lay claim to having made a discovery, the very soul of science.

Behind Mach's claim, potentially so destructive of science, lay his indebtedness to Spencer's notion of knowledge as a purely biological adaptive process,[49] within which the progress of the mind consisted in the assimilation by the organism of ever more differentiated aspects of sensations. Spencer articulated his topic, as he did countless others, with a fluency which helped him to make even the most complicated topics appear dazzlingly simple and earned him the aura of profound greatness. That aura gave the impression to many—Darwin was one of them[50]—that their reading of Spencer made them commune with one of the greatest intellects in history. They did not suspect that Spencer merely mesmerized their philosophical incompetence by his own fashionability which was to fade soon as do all fashions. Philosophical authors, who sway their own generation but are almost completely ignored by the next, should not, however, be entirely forgotten. They are extremely informative about the ease with which obvious problems can be overlooked in a system which is the rage of the day.

This oversight is to occur all the more naturally when one has already been treated to genuine insights which are not lacking in Spencer's 'Genesis of Science' first published in 1854.[51] It must have been refreshing to hear in those days that scientific thinking is not something essentially different from commonsense reasoning, that all scientific knowledge was ordinary, that is, qualitative in its primordial form.[52] A historian of science can only applaud on finding Spencer declare that the understanding of the original stage is the key to the interpretation of all subsequent development. It is in fact more than rhetoric to speak of the embryology of science[53] and point out that disregard of that embryology leads to artificial classification of the sciences, such as was offered by Oken, Hegel, and Comte. Even a few salient data about the early phases of modern science were enough in Spencer's hands to discredit Comte's claim that the systematic organization of science in decreasing order of generality was also the order of their historic development. The claim as Spencer put it was metaphysical and thus contradicted the positivist Comte.[54] It was even more insightful to hear from Spencer in an age mesmerized both by the deductionism of idealists (Hegel) and by the inductionism of empiricists (Mill) that analysis and synthesis are always in inseparable interplay.[55]

That sound message was not Spencer proper. In fact disaster was in sight as soon as he entered the scene with his own brand of psychologism. He now claimed that the psychological development of scientific thought was also its historic development.[56] Here was, as so often in the past, the philosopher playing the role of the historian without having developed enough respect for the hard facts of history. Those familiar with the sweep and ease of Spencer's pen will not need details about the true merits of those twenty or so pages in which he moved from the aborigines' perception of likeness among their animals to the determination by Babylonian magi of the planetary cycle known as Saros. What needs to be recalled are some obvious consequences if his claim was true that psychological development (as imagined by him) was equivalent to historical development. The chief consequence of this thesis, which Spencer himself spelled out as the conclusion of his long essay, was the substantial parallelism of development. This meant that any ancient culture had to show essentially the same progress in science. Actually, some of the illustrations used by Spencer could but appear counter-productive on a little reflection. In reconstructing psychological development nothing was more tempting than to relate the origin of counting to reliance on one's ten fingers. But the decimal counting was not at all universal in primitive and ancient times, nor was the vigesimal system that could be traced again to man's fingers and toes. Again, while vertical sticks were used in many cultures (primitive gnomon) to estimate angles from the length of the shadows they cast,[57] the geometries of which ancient cultures could boast were indeed very different in development.

The fact that only the Greeks raised geometry to the level of science was a resounding rebuttal of the substantial parallelism advocated by Spencer. Nor was it permitted on the basis of parallelism to jump, as Spencer did, from the Greeks to Galileo. That in between there were some Christian centuries steeped in the Gospel was hardly to be noticed within Spencer's perspective. His shortsightedness on such and similar points was strengthened by the fact that his admittedly main source of information on the history of science was a book which in spite of the renown it enjoyed in its day did not do much justice to that history. The book, William Whewell's *History of the Inductive Sciences*,[58] is relevant here not only because of its having been relied upon by Spencer and many others as the last word on the topic, but also and mainly because it provided no role for the Word speaking from the Gospel in intellectual history. Whewell's failure to do so should seem all the more revealing because he was in holy orders and he was firmly convinced that natural and philosophical science and historical reflections on it lead to a recognition of the God of nature.[59] But Whewell is also an example of those many Christian thinkers who are unaware of the fact that on the basis of their philosophies it is impossible to argue effectively on behalf of God and to put into focus the role of the Gospel in intellectual history. The same philosophies can also be shown as being incapable of doing justice to the history of man's science about nature and to that very science itself.

Whewell's was a strange mixture of Baconian empiricism

and Leibnizian apriorism, a mixture which lacks ingredients needed for carrying out any of the tasks just listed. Worse even, he did no original studies in the history of science. On the basis of that mixture he merely articulated clichés already hallowed in that field. Thus Whewell dismissed ancient cultures as foreign, that is, irrelevant to the history of science, and praised the peculiarity of the European mind.[60] Like many before him he did not care to ask why that European mind was so different. In accord with a long-standing custom he jumped from the Greeks to Galileo; his long chapter on the medievals was a mere abuse of them. About Adelard of Bath and Grosseteste he mentioned only their names[61] in clear evidence that he did not read what they wrote. The whole medieval period was for him a 'mid-day slumber',[62] a metaphor which makes sense only if there had been true alertness in the morning. The historical situation demanded, however, more than pleasing metaphors. That the morning did not pass into a noon diffused with sunshine was more a problem connected with the Greeks who rose with the sun, than with the medievals allegedly snoozing in bright noon. Whewell's answer for the failure of the Greeks was the basic tenet of his philosophy of science, namely, that facts of observation must be matched with distinct ideas homogeneous with those facts if the latter are to generate science.[63]

The tenet was effective only in generating tautology. Clearly, it remained to be answered how ideas homogeneous with facts are being generated. A historian of science blessed with hindsight can, of course, list the ideas that were found 'homogeneous' with facts and therefore constituted a discovery. But this still leaves unexplained the act of discovery as an act of perceiving truth embedded in facts. While the historian of science is not called upon to explain the act of discovery, he can at least identify the sources of some great intellectual changes that ultimately made possible the sequence of discovery which corresponds to a viable birth of science, a birth which science did not have among the Greeks of old. As the next lecture will show, the historian who achieved that fact did so by keeping his eyes open to the light which the Gospel brought into intellectual history. The feat also meant at least a theoretical end to a historiography of science culminating in Whewell's *History*. It was a historiography in which all other cultures except the European were uninformative rubbish with respect to science and in which no question could logically be raised about the distinct origin of a science which is indeed European.

In that historiography science arose as a Deus ex machina, with Bacon having been assigned the rather unscientific role of playing God.[64] He certainly failed to inspire the writing of a genuine history of science although he advocated it in memorable words gracing as a motto Whewell's *History*: 'A just story of learning, containing the antiquities and originals of knowledges, and their sects; their inventions, their diverse administrations, and managings; their flourishings, their oppositions, decays, depressions, oblivions, removes; with the causes and occasions of them, and all other events concerning learning throughout the ages of the world.' Almost three hundred years after Bacon it was still true that, as he put it, such a just story is 'I may truly affirm to be wanting'.[65] It failed to be written because due in no small part to the persuasiveness of his empiricist precepts no proper attention could be focused on the origin of a history which shows the creative work of the mind within the empirical. It is precisely because of that creative work of the mind that the empirical, be it the datum of science or of art, is raised above the level of uninformative rubbish and turns into history.

THE FASHION OF STRANGE MOULDS

About William Whewell, whose *History of the Inductive Sciences* had since its publication in 1837 set a standard for over half a century, it was noted that science was his forte and omniscience his foible.[1] During the heyday of Whewell's *History* and for some time afterwards, the history of science was cultivated often by those who had science for their speciality while its history was treated by them with an ease reminiscent of omniscience.[2] The major break with this state of affairs came when expertise in science was combined with a meticulous study of its history, a feat certainly unusual, but no usual a scholar was Pierre Duhem who made that break with a rather unscholarly past. That he made history in the cultivation of the history of science is, as is all too often the case with such breaks, a matter of restrospective insight. It received a forceful expression when in 1961, at the Oxford symposium on the history of science, Henry Guerlac referred to Duhem as the 'acknowledged teacher of us all', the one who showed historians of science how to 'focus upon the evolution of key scientific ideas and concepts'.[3]

About that remark one can only say that no remark was more well founded, but also more overdue, and unfortunately, less appreciated. Had Guerlac voiced not only the truth of Duhem's pre-eminence, but also the truth of a widely shared conviction about Duhem, an international congress on him would have most likely taken place in 1966, the fiftieth anniversary of his untimely death at the age of fifty-five. While in 1966 there was a rash of activity (including an international congress) and a fresh flow of publications to celebrate the fiftieth anniversary of the death of Ernst Mach, hardly a word was said or printed about Pierre Duhem, except perhaps in such a way as to enhance the stature of Mach.

Obviously, it was not true that Duhem was acknowledged as a teacher by all or most historians of science. Indeed, the few publications that give some though meagre information about his work in the history of science were not written by historians of science.[4] The teacher they celebrated was rather Ernst Mach, a rather unreliable teacher as a historian of science. Indeed, once the euphoria of the Mach-anniversary was over, there appeared several important publications, each a challenge to the myth about Mach's pre-eminence as a historian of science and to his importance in the history of modern science.[5] Duhem is still to be acknowledged as a physicist far more creative than Mach, and as the one who made possible a genuine historiography of science, the rise of which Mach's *Science of Mechanics* only helped to postpone. Such a historiography is free of the dubious logic of bright mornings followed by dusk at noon, and is not saddled with the acrobatics of jumping a millennium from the Greeks to Galileo. The safety of historiography ushered in by Duhem derives from its chapter about the medieval origin of science, a chapter which shed a coherent light on the whole development of science in the sense in which Spencer wanted to have an embriology of it.

What is best known and also universally acknowledged about Duhem's feat is its massive documentation. The fourteen hundred pages of his three-volume study of Leonardo da Vinci and the five thousand pages of his *Système du monde* would represent unique scholarship even if their massive documentation had been taken from printed sources alone. But a considerable part of that vast material was taken from medieval manuscripts and many of them were tracked down, deciphered, studied, and excerpted by Duhem, who had no xerox machines, not even ballpoint pens, to expedite his research. He worked alone, a fact almost impossible to believe in view of his gigantic output, invariably of highest scholarship.[6] The rumour could therefore naturally arise that Duhem had been assisted by a research team, a rumour possibly fed by the well-known assistance which a Needham and a Thorndike needed and received from highly trained collaborators.

Beneath Duhem's colossal work demanding a single-minded commitment there must have been a specific motivation, a point easy to verify in his case, though it is a point not to the liking of many. Duhem, educated at the Collège Stanislas in Paris, was pre-eminent in letters as well as in the sciences. By entering the École Normale he decided to devote his life entirely to research in theoretical physics. The essay he produced within three years for his thesis was brilliant to the point of undermining some favourite theorems of Maurice Berthelot, the leading French physical chemist of the day, and, as minister of public instruction, the undisputed ruler of the French academic establishment. Berthelot not only had Duhem's thesis rejected,[7] but also made it clear that 'this young man [Duhem] shall never teach in Paris'.[8] Berthelot's reaction undoubtedly had much to do not only with his deeply hurt professional pride but also with his militant anticlericalism.[9] A Duhem, who made no secret of his deep Catholic convictions,[10] could only be a thorn in Berthelot's side.[11] So it happened that for the next thirty years Duhem, one of the greatest French intellects around the turn of the century, was denied a chair in Paris, the only place where he truly felt at home. It was by rushing back and forth between the university of Bordeaux and the Bibliothèque Nationale in Paris that Duhem did his epoch-making research in medieval science.

As usually happens, a firm commitment grows even stronger under fire and this is what happened to Duhem. He saw clearly that under specious references to the rule of reason, the French academic leadership of his time carried on an unremitting campaign against Christianity and the Catholic Church. A hundred years after its publication, Condorcet's *Sketch* of cultural history was more than ever the secular gospel with its crusading spirit in undiminished vigour. A chief slogan of that crusade was the alleged incompatibility between Gospel and science, with the Catholic Church being the chief instrument of perpetuating medieval darkness. It is safe to assume that when a scholarly biography of Duhem, based in part on his hitherto unpublished correspondence, is written, it will become amply documented that Duhem's turning around 1900 to the history of science, and especially toward medieval science, had its motivation in his feeling of being called to arms in defence of Gospel and Church. At any rate, he voiced a life-long conviction when in an already published letter of his, written in 1911, he called a plain lie the widely shared belief about a basic opposition between science and Church. He did so with an implicit reference to his reconstruction of the importance of Buridan and Oresme,[12] in clear evidence that he saw his research in the history of science as a most needed and most effective apologetics.

Apologetics, Christian or other, is always a risky enterprise. Once caught in apologetic zeal, one can, especially if one is a historian, easily overshoot the target. Too much defence can easily leave one wide open to attacks and even make one quite defenceless. It is indeed possible to argue that in some particulars Duhem attributed too much to his medieval heroes, to Buridan, Oresme, and to others. But the evidence he brought to light is also too much to be taken lightly. He certainly proved that in respect to science the medievals were not caught in noonday slumber, or that mental stupor in which Whewell believed them to be lost. On the contrary, they opened up crucially new perspectives useful for science. That these new perspectives are most helpful for the historiography of science as well as for Christian apologetics may disturb only those who cultivate that historiography in the hope that it may provide ammunition for anti-Christian apologetics, however covertly cultivated.

Those who prided themselves on being historians of science shortly before Duhem turned to the medieval manuscripts might have been expected to rush to the sources after the printing in the 1880s of Leonardo's notebooks which showed that as a scientist Leonardo owed much to medieval sources. Neither those notebooks nor their sources had attracted at first any leader among historians of science. Obviously, more was needed in the way of motivation than love of facts, a love to which lip-service is paid more often than true homage is rendered. In Duhem's case it was indeed a deep love of the Gospel that made him shift part of his interest to medieval science. The shift must have appeared a senseless fragmentation of energy on the part of a Duhem who even after having made his mark as a historian of science never considered himself to be one, and who, rumours to the contrary, would certainly not have accepted the chair for the history of science at the Collège de France had it been offered to him. To the end he considered himself a physicist and kept telling friends: 'I shall either return to Paris as a physicist or I shall never return there'.[13]

To steal many precious hours from physics demanded on

Duhem's part a motivation, his love of the Gospel, a contention which will not appear far-fetched once the intensity and sincerity of his spiritual life is recalled. If, however, the the love of the Gospel can be so rich in heuristic value as to issue in scholarly results evidenced in Duhem's massive volumes, belief in the Gospel will not perhaps appear something to be frowned upon by the historian of science without further ado, an attitude which at any rate is not characteristic of objective reasoning. That objectivity is glaringly lacking in remarks in which interest in medieval science is dismissed as something in which mostly Catholics are interested. The real issue, however, is the fact that so many non-Catholic historians of science fail to be appreciative of the facts of medieval science although they claim to be motivated only by reverence for facts regardless of their provenance. Behind their slighting of those facts there lurks time and again a familiarity with Duhem's books which is not better than Duhem's identification in an American publication as professor at the 'University of Cabrespine',[14] a picturesque hamlet not far from Carcassonne but as removed as the latter is from the academic world.

Among those facts Duhem singled out one as being of uppermost and decisive importance for the whole future of science which was still to see its viable birth. The fact was the Christian refusal to accept the ancient pagan dogma of the divinity of the heavenly regions and bodies. While it is true that some Greeks, notably Democritus and Anaxagoras, held all bodies to be of the same material nature, the foremost Greek thinkers from Plato to Ptolemy firmly upheld the divinity of the heavens. In that crucial respect the Greek genius was unable to break a pattern characteristic of all ancient cultures. It failed in its scientific endeavour precisely because of the mesmerizing impact of a divine sky determining everything on earth according to its perennial revolutions. Within the perspective of that cyclic determinism everything on earth, even the processes of history, could but turn into the treadmill of a vicious circle. In that outlook it was impossible to decompose the reality of motion into impetus and inertia and thereby lay the foundations of classical physics. Equally important, the spectre of that treadmill made it impossible to muster confidence in progress based on the investigation of nature.

For man to free himself of the monstrous conception of a divine realm of celestial bodies ruling all processes on earth, physical as well as human, man needed the perspectives of Christianity. For as Duhem illustrated with words from their own mouths, the Greeks, the Arabs, and even some of the most illustrious rabbis from Philo to Maimonides, fell prey to that cosmic vision of eternal cycles, nay, they all were busy in justifying it with arguments that were scientific only in appearance. Such was the famous and emphatic conclusion reached by Duhem as he completed his magisterial analysis of Greek, Arabic, and Jewish astronomical lore.[15] The conclusion saw print in the second volume of his *Système du monde*, a point which has a relevance to be noted later. The sixth volume of the *Système* contained another famous statement of Duhem, a statement almost invariably taken out of context by his critics and detractors. The statement is his identification of the moment of the birth of modern science with the condemnation of scores of Averroist theses by Étienne Tempier, bishop of Paris, on March 7, 1277.[16] Clearly a Duhem, who was a firm advocate of the idea that science advanced at an almost imperceptibly slow pace, did not mean to be taken literally. He merely saw in that condemnation a graphic instance of that long struggle which Christians had to undertake often with perplexity against the doctrine of eternal cycles, a doctrine most alluring but also incompatible with their belief that the hero of the Gospel was not only the Word made flesh at a unique point in history, but also that very Word in whom the Father created all *in the beginning*, the ultimate origin of all history.

As a Christian, Duhem knew very well that whereas the dogmas of Creation and Incarnation meant, theoretically speaking, an absolute and most revolutionary break with a past steeped in paganism, the articulation of those dogmas in time and their impact on history was an uphill fight never to be completed. Herein lay a concept of intellectual revolution which, if studied and heeded, might have saved the recently born historiography of science from getting sidetracked in the conceptual quagmire of mutually destructive revolutions pre-empting the notion of scientific progress. The scientific revolution permissible within Duhem's perspective was quiet and slow in asserting itself but assuring a progress immune to the débâcles of revolutions. Duhem would have relished Whitehead's famous remark that it is doubtful that as great a thing has happened with so little stir ever since a babe was born in the manger.[17] The stir was quietly steady in both cases, and full with commitment to a coherent physical and historical reality.

This commitment to reality was for Duhem, the historian and philosopher of science, a tenet never to be abandoned or questioned. Here too he has been rudely misinterpreted, especially when described as a chief ally of Mach, for whom the analysis of the history of science showed that science is purely operational in the sense of being an economical correlation of sensations with no information about reality as such. True, Duhem was the author of a monograph which came to the conclusion that 'despite Kepler and Galileo we believe today with Osiander and Bellarmine that the hypotheses of physics are mere mathematical contrivances, for the purpose of saving the phenomena'. It is, however, hardly a sign of scholarship to forget that this phrase was immediately followed by another which brought his monograph to a close and in which Duhem declared: 'But thanks to Kepler and Galileo, we now require that they [the hypotheses of physics] save *all* the phenomena of the inanimate universe *together*.'[18]

Unlike Mach, an agnostic turned Buddhist,[19] Duhem the Christian was firmly committed to the reality of the universe, that is, to the totality of consistently interacting things, the only kind of totality that could come from the hands of a fully rational Creator. It is that totality which is the basis of Einstein's general relativity and of truly scientific cosmology of which his relativity is an indispensable basis. Mach, on the contrary, sidestepped cosmology in the very sight of a most central issue of it,[20] and grew increasingly suspicious of the absolutist traits and cosmological relevance of Einstein's train of thought.[21] Mach was resolved not to let science touch on reality as such because that would have made science an ally of all rational believers in God who always rest their case on the reality of the universe in so far as it points to God through its peculiar coherence proclaiming its contingency. Duhem, the Christian, could only be appreciative of such a universe and this is why he held that the gradual refinement and universalisation of physical theories provided an ever more reliable picture of its ontological reality.[22]

What seems to be in full evidence here is an intimate epistemological connection between cosmic and intellectual history, anchored in a notion of ultimate origin, a connection that would have been unacceptable to Mach. Indeed, for all their protestations of mutual esteem,[23] Duhem and Mach were poles apart. The satisfaction they felt on using the same method, historical investigation, to achieve the same immediate aim, the overthrow of the naïve realism of classical physics, prevented them from airing the very substantial differences between their real objectives and between their basic premises. Mach, for instance, acknowledged his debt to Duhem in some particulars concerning Galileo's forerunners, without even hinting at Duhem's assertion of a medieval, Christian origin of modern science.[24] Duhem, in turn, wrote in 1903 a review of the French translation of Mach's *Science of Mechanics* without exposing its anti-Christian thrust.[25] Mach, who was able to conceal for years his parting with Einstein,[26] would probably have kept silent had he learned about the contents of Duhem's letter to the rector of the newly established Institut Catholique in Paris. In that letter Duhem not only decried the staple misrepresentation of the Middle Ages, but also suggested the establishment at the Institut of two chairs, one for the philosophy, another for the history of science, and for a purpose which Mach would have certainly classed as a most disreputable form of Christian apologetics. The purpose in question was conveyed in Duhem's insistence that the analysis of scientific theories and the presentation of a genuine history of them were most effective means of countering the hallowed claim that Gospel and science were irreconcilable.[27] Far from being in a real sense in the camp of Mach, Duhem, the historian and philosopher of science, was a Christian apologist precisely because his Christian faith liberated him from fears of keeping constantly in sight the perspective of ultimate origin.

Such a role, the role of a Christian apologist, is a role of which no one should be less ashamed than a Fremantle

Lecturer. At any rate, the role of apologist will appear objectionable only to those who believe in the possibility of thinking and writing for no purpose. A Turgot, a Condorcet, a Comte, a Spencer, a Mach, were at least as fervent apologists as was a Duhem, a truth valid even of those historians of science whose apologetics of pragmatism demand a systematic slighting of him and often the covering of him with silence. Those historians of science, who with Dilthey cheerfully registered the relativization of all ideas throughout history,[28] are still to show how any particular truth can be saved from being dissolved in the subjective moulds of pragmatism unless that truth is connected somehow with absolute truth, the kind of truth which alone had appeal for Duhem.

There is no great distance from that pragmatism to that diffuse universalism which through the influence of George Sarton served as the first fashionable mould for the historiography of science in this century. It was that mould that kept together the countless data in Sarton's chief contribution to the historiography of science, the five massive volumes of his *Introduction to the History of Science*.[29] They are indeed at best an introduction but certainly not a history. They could not have become an account of history even if Sarton had intended to present his material in such a way. The account in question must be more than a recital of any and all facts. It must be equivalent to a judicious choice among facts and an interpretation of them, a casting of them into a mould suggested by that sequence of theirs which reveals their relative importance. That sequence indicates that not all ancient cultures were equally important for the scientific enterprise and that the enterprise itself failed to go beyond a certain point in each of them, a truth which holds even of Greek science. This latter point is important to make because Sarton's only work which is a history of science is a two-volume account of the history of science in ancient Greece. It was his last work completed shortly before his death in 1956. He would certainly have been hurt had he lived long enough to read L. Pearce Williams' statement in the *British Journal for the Philosophy of Science* that in Sarton's *History of Science* the actors are either 'good guys' or 'bad guys' and from their clash comes the dynamics of history.[30]

Even the 'good guys' and 'bad guys' distinction would not have been an altogether useless mould for the interpretation of history had it been used with the willingness to admit that as far as science was concerned 'good guys' and 'bad guys', that is, creative and non-creative minds, were not present in the same proportion in all cultures. But Sarton's 'liberal' spirit could not admit that kind of discrimination. He saw the dawn of science everywhere: in the stone-age as well as amidst today's aborigines, nay in every child's awakening mind.[31] This approach to the dawn of science might have helped the anthropologist though hardly the historian of science whatever their common need of Sarton's professed largeness of mind. With that largeness Sarton might have been expected to find many a 'good guy' during the long medieval centuries. Were not those centuries Sarton's professed speciality?[32]

Whatever the merit of Sarton's scholarship as a medievalist, he certainly saw the medievals through very special glasses. A subtle disclosure about the distortion of those glasses came in the very last phrase of his last book which carried the history of science to the end of the first century B.C. 'The appearance', reads that phrase, 'of a new incomprehensible mystery, that of Jesus Christ, and its gradual triumph characterized an entirely new period'.[33] By that new period Sarton obviously meant the medieval centuries, though his reference to Jesus Christ as an 'incomprehensible mystery' was a cryptic clue as to whether Jesus and his followers would be among the 'bad guys' or the 'good guys' as far as science is concerned. Sarton's previous publications had already made it clear that he counted Jesus and his followers among 'bad guys' uninterested in the pursuit of knowledge. Sarton made this abundantly clear in 1935 in the third of his Culver Lectures delivered at Brown University. That lecture dealt tellingly enough, under the title 'East and West',[34] with that universal or ecumenical spirit which for Sarton was the mould into which the facts of history had to be poured in order to become intelligible. For Sarton, the historian of science, East and West made equally valuable contributions to science, a contention which should reveal to anyone familiar with the rise of science in the West that its juxtaposition with the East in a so-called ecumenical spirit is indeed a strange mould.

That in spite of its obvious inappropriateness to accommodate the salient facts of scientific history the mould in question had great appeal for Sarton, the avid collector of facts, must then have had a strong motivation. Its source was Sarton's 'faith of a humanist' which he set forth as early as 1920 in his capacity as the editor of *Isis*, half a century afterwards still the leading journal in the field. Through the glasses of that faith Jesus Christ could at best appear as an 'incomprehensible mystery', not, of course, in the sense in which His uniqueness raises Him above the ordinary moulds of comprehension. Whatever the incomprehensibility of Jesus, Sarton's incomprehension of Him was strange indeed. Contrary to Sarton, Christ's emphasis on love and humility did not mean that there was only one side of the spiritual currency which He wanted to make universal, namely 'that knowledge without charity is not only useless but pernicious; it can but lead to pride and damnation'.[35] Jesus, who used a coin graced with reference to God on one side and to Caesar on the other to thwart efforts to trap Him in one-sidedness, cannot, without doing grave injustice to His message, be pictured as one not inspiring a charity fully permeated with knowledge. Was He presenting His message only with charity and not also with a knowledge that left His opponents speechless time and again?

Sarton's ecumenical mould, which turned strangely unecumenical when it came to Jesus, gave itself away even more when he declared that 'development of Christianity was a first attempt to bring together the Hebrew and the Greek spirits, but as the Roman Christians hardly understood the former and misunderstood the latter thoroughly, the attempt was an utter failure'.[36] This phrase was not better than the rhetoric of Turgot, Condorcet, Comte, Spencer, and Mach. Christians had remained an utter failure for them as well for Sarton, a self-declared historian of medieval science. While Sarton acknowledged the rise of experimental method among medieval Christians, he credited it to their having been the disciples of the Arabs. For Sarton the 'first deliberate vindicator of experimental spirit'[37] was Leonardo da Vinci and he made much of the alleged stifling of Jewish talents in Christian medieval ambience. That he saw in medieval scholasticism only a barren exercise of the intellect will make complete that rudely unecumenical appreciation of ages which he did not call dark but painted certainly in dark colours.

Clearly, the glasses Sarton wore were ground more according to a long outmoded interpretation of cultural history than according to some crucial findings about it, findings of which he could not be unaware. Sarton, who wrote an enthusiastic review of the first volume of Duhem's *Système du monde* and expressed his eagerness to see the second and further volumes,[38] suddenly fell silent on Duhem. Those familiar with the Freemason Sarton's deep-seated anticlericalism, coupled with his dogmatist socialism born out of a youthful cavorting with Marxism, will easily understand that Sarton could only feel dismayed on reading in the second volume of Duhem's masterpiece that Christianity was needed to discredit and jettison the monstrous doctrine of the Great Year and open thereby the true perspectives needed for the still-to-be-born scientific enterprise. Unable to cope with Duhem on the level of medieval scholarship Sarton was left with the alternative of ignoring him as if erudition and publications did not count. Neither the subsequent volumes of Duhem's great work nor his three-volume work on Leonardo were reviewed by Sarton, who for many years wrote most reviews for *Isis*. Duhem's death was not recorded in *Isis* when it resumed publication after World War I, nor was he commemorated at the twenty-fifth anniversary of his death in 1941. Most importantly, although in 1937 Sarton called on the readers of *Isis* to support the eventual publication of the second five volumes of Duhem's *Système du monde* still in manuscript, he did not refer to Duhem's two great works in his *Introduction* in which he should have mentioned them in many a context, all the more as he found place there for many insignificant items. Sarton's stance toward Duhem did not measure up even to a low-quality ecumenicity of spirit, let alone to the standards of good historiography. A chief result was Sarton's exercise in equivocation about the origin of science, an exercise pre-empting the science of its origin. With the question of the origin of science being side-tracked, the historiography of science could but be derailed into that alluring terrain where a so-called ecumenism makes one insensitive to some crucial differences between East and West and where everything will appear clear except the fact that science as a self-sustaining enterprise, the only science worthy of that name, appeared not in the East but in the West and in a West which was unmistakably Christian.

That the mould of that ecumenism can be only a Procrustean bed for scientific history has ample illustration also in Joseph Needham's celebrated work on science in China, a work as voluminous as that of Sarton, equally informative in countless small details, but also equally effective in helping obliterate the unique grandeur of the Christian West. From 'The Unity of Science: Asia's Indispensable Contribution' (Needham's first major postwar address) to 'The Roles of Europe and China in the Evolution of Ecumenical Science' (1966; the last of his essays collected in a huge volume)[39] Needham kept preaching a gospel and he still preaches it. His gospel is so ecumenical as to make Christ appear a very thin, almost invisible figure. Needham himself publicly stated that his is an adherence to a Christian faith best articulated by Tyrrell, for whose modernism he could not give high enough praise.[40] Few in his crowded audience were aware of the fact that no modernist kept faith in Christ, that is, a faith in which Christ remained what He had claimed himself to be: *the* road, *the* life, and *the* truth. The obliteration of Christ's uniqueness can but invite the turning of Christianity into one of the many factors of history, all on equal footing, all of equal merit within an ecumenism that can accommodate everything except a Christianity anchored in the uniqueness of Christ. Such is the deeper background for that simile or mould, titration, a favourite with Needham,[41] for the explanation of the rise of science.

To fall back on chemistry for a mould to explain the

rise of modern science, a crucial aspect of human history, should seem suspect to anyone still mindful of the fact that the mind is not a machine, mechanical or electronic, and much less a mere chemical test tube. Titration as a mould is also misplaced because unlike the experiments of chemistry, the experiments of history are not repeatable. Furthermore, since titration is a careful mixing of various ingredients to produce a solution of required strength, one could but expect Needham to offer a properly titrated bibliography.[42] He certainly offered an immense one, but an entry, representing massive scholarship, is strangely missing there. That entry should have contained the two major works of Pierre Duhem, which certainly would have merited being cited far more than many a work quoted and praised by Needham. Like Sarton, Needham too had the personal privilege to disagree with Duhem's interpretation of history, an interpretation unabashedly Christian and irreconcilable with that ecumenism which Sarton and Needham held high. But as historians of science, claiming a high measure of scholarship, neither Sarton nor Needham had a right to be silent on Duhem, whose scholarship was in the end acknowledged even by Berthelot,[43] though not to the extent of letting the greatest French scholar of the day return to Paris and dominate the scene there with his visible presence.

The figure of Duhem is clearly present, though only as a figure to be turned into a shadow, behind the notion of revolution that is possibly the most fashionable conceptual mould into which the history of science is being cast nowadays. Alexandre Koyré, the prime source of that fashion, had been dead for a decade when in the early 1970s he was referred to as the 'dean of historians of the scientific revolution'.[44] It was then that the notion of revolution as the chief paradigm or mould of the history of science reached the peak of its fashionability. The process was not, however, free of that characteristic which is the inability to account for fashions in a rational way. Rationality was hardly given its due when Thomas Kuhn, author of *The Structure of Scientific Revolutions*, was credited with deep insights into the nature of the scientific endeavour, while in the same breath he was pointedly asked to 'stop populating science with new entities [by reifying mere concepts or paradigms] so that we may more easily have access to his insights'.[45] The merits of Kuhn's book could but appear highly questionable when it was declared to be possibly that book in this century on the history of science which Mach's *Science of Mechanics* was a century ago, but when in the same context it was also shown that one can easily take any set of propositions in Mach's book and prove them invariably wrong.[46] Again, what was the measure of consistency, without which there is no rationality, when in 1965 Kuhn was declared to be one of the outstanding philosophers of our time by that very analyst of the *Structure* who found in it more than twenty different meanings of its key word, the paradigm?[47]

Those unimpressed by the mythical clarity of revolutions are not surprised to find that dwelling on revolutions is hardly a road to clarity. The idea of continual revolutions advocated by Mao and his captives may not be as dead as some would like to believe, but there can be no doubt about the dismal confusion into which eight hundred million Chinese have been thrown through a systematically popularized emphasis on keeping up with revolutions. Revolutions are also well known for devouring their chief advocates and this sad though well-merited fate now seems to have caught up with Kuhnians, rapidly multiplying the number of revolutions in science. Except for intellectual Maoists with masochistic delight in revolutions, the revolutions will have lost their appeal and more importantly their meaning, if they are turned into daily fare and affair. Rational discourse is not promoted when revolutions mean normalcy and when one is therefore left with normalcy meaning revolution.

Of this confusing state of affairs little could be suspected when Koyré published his *Études galiléennes* in 1939. On a cursory reading its argument looked like the turning of those in Plato's cave toward the full brilliance of a sun never to be overcast again. But the signs of permanently turbulent weather were not difficult to spot in the *Études*. Koyré himself noted that the early twentieth century witnessed a scientific revolution just as decisive as the revolution implemented by Galileo and Descartes.[48] Plato himself never spoke of repeated turnings toward the sun, let alone of ever new and ever better suns casting forth the rays of permanent truths. In fact, Koyré's first clarification on the first page of the *Études* came not from Plato but from Gaston Bachelard, from his anything but Platonist interpretation of the progress and dynamics of scientific thought.[49]

Actually, there could be no progress of science if Bachelard was right in claiming that the only true philosophy was a continual oscillation between rationalism and realism by which he meant an oscillation between idealism and empiricism. Realism, as a median position between the two, he could not have had in mind because he jettisoned the only realism that can be a real median position. It is inseparable from the reality of substances to which Bachelard gave good riddance.[50] About substances it is well to recall that they are not entities of observation but entities of postulation. They were postulated by Aristotle and by all those after him who did not want to despair of some permanence and continuity in the realm of the really experienced. Securing genuine continuity to the scientific enterprise was not apparently a concern with Bachelard, more given to poetical language and far-fetched metaphors than to consistent reasoning. Rationality and consistency, without which there is not continuity and permanence, were at stake when Bachelard called for a 'theory of the objective in defiance of the object' and for a 'thinking in defiance of the brain'.[51] It should not therefore seem surprising that he saw contemporary science as a scene of 'most decisive psychological mutations'.[52] While even in biology, where mutations have been intensely studied and measured, their range and role are not wholly understood, a reliance on obviously unspecified psychological mutations can but make shambles of any field of study, a result for which the uncritical enthusiasm for Koyré's *Études* is largely responsible.

The starting point could not have appeared less innocuous and more scholarly. It was also indicative of a brilliant tactician's skill of damning with faint praise. Koyré's *Études* was in fact introduced with the statement that in view of the magisterial work of Duhem it was superfluous to note the fruitfulness of historical studies for philosophical reflection.[53] But the very same introduction reached its climax with the declaration that 'classical physics, as a fruit of the thought of Bruno, Galileo, and Descartes, does not continue—Duhem's emphasis on historical continuity notwithstanding,—the medieval physics of the Parisian precursors of Galileo. It is situated on a very different plane, a plane which we would like to call Archimedean. In fact the precursor and magister of classical physics is not Buridan or Nicole Oresme, but Archimedes'.[54]

With that thesis of Koyré, in whom many younger

historians of science found their master and inspiration, the
silence about Duhem, already made fashionable by Sarton,
was turned into a verdict of guilty of revisionist miscon-
struction. With such a verdict on Duhem his massive
volumes could be safely left to gather dust. Koyré, who in
spite of his professed ecumenism of spirit, could not muster
sympathy for Pascal,[55] could hardly feel appreciation for
Duhem who knew Pascal's *Pensées* by heart[56] and who like
Pascal knew what it means to believe in, not the elusive
God of Plato, but the God of Abraham, Isaac, and Jacob,
the only God that is believed to have entered history, social
as well as intellectual.

Koyré's juxtaposition of Bruno, Descartes, and Galileo
was somewhat hasty. Twenty years later his reading of
Bruno convinced Koyré that Bruno's was not really a
scientific mind.[57] This was probably the most charitable
thing that could be said of Bruno, who decried the file of
geometry as detrimental to scientific reasoning.[58] Clearly,
Bruno was worlds removed from Galileo and Descartes as
well as from Archimedes. Koyré's error about Archimedes
could have been spotted even easier than his blunder about
Bruno. The latter's contempt for the file, or cutting edge,
of geometry remained unnoticed by the few who in recent
years had the courage and scholarship to delve into Bruno's
obscurantist volumes.[59] As to Archimedes, one only needs
to page through the single volume of the English translation
by Sir Thomas Heath of all his works[60] in order to see that
although Archimedes speaks much of geometry, indeed he
speaks of nothing but geometry, he nowhere refers to the
application of geometry to the analysis of motion. That
analysis, the birth of science about a world in motion, had
started with the Parisian precursors of Galileo as amply
documented by Duhem.

They became the starting point because they were imbued with what is Gospel truth for Christians though it had never been for the Greeks of old, namely, that the universe is not God but only the fully consistent artifact of a rational Creator. Because of their belief in that consistency, they could approach with quantitative eyes the phenomenon of motion in a broad sense, an approach alien to the Greeks. It remained alien to them in the measure in which Plato's utterance, 'God does but geometry', proved not to be a prompting for anyone among them to come to grips with concrete earthly reality always in uneven motion. That the medievals made much of such utterances originating in classical antiquity and that in particular they eagerly seized on Archimedes' works was due to their worshipful respect for the utterance, 'You have disposed everything in measure, number and weight'. This phrase, one of the most often quoted phrases throughout the Middle Ages,[61] was not predicated on a demiurge but on a Creator. It should not therefore be a cause of astonishment that the time-squared law of free fall was first stated not in Archimedes' works but in a scholastic context which set the tone of Galileo's early writings.[62] Unlike Archimedes, Galileo took the human mind for the Creator's finest product, created to His very image and therefore equipped to fathom at least in a quantitative way all His other products.[63] In that respect too Galileo was on a continuum with the medievals. Where he was not on a continuum with them was his proverbial hybris that made him blind, among other things, to the limitations of the quantitative method. For that blindness the highest price was to be paid by that very science which, some time after its first and only viable birth, mistook the midwife, the quantitative method, for the womb or the Gospel truth that man, the master and possessor of nature, was not a mirror image of quantities, Platonic or worse, but was created to the image of God.

The interpretation of that science is today the prey of many a historian with no fondness for the Gospel. They cast science in moulds which are so strange to science as to deprive it even of the possibility of progress. Some of those moulds are plainly disreputable, such as the one in which the rise of science in the West is ascribed to the spread of hedonism following the Renaissance, and the stillbirth of science in China is traced to sexual repression.[64] Though not disreputable in that sense, not much more reliable is the mould of Darwinian natural selection championed by some historians and philosophers of science who picture scientific ideas engaged in a struggle for survival.[65] They almost invariably forget Darwin's warning that an evolution based on natural selection forbids its student to use the words lower and higher.[66] Once this warning of Darwin is remembered—he himself mostly forgot about it—the progress of science as a clash of concepts with greater and smaller survival value will appear a shabby affair indeed. Its incoherence hardly reflects those intellectual heights to which Condorcet saw mankind being led through science. But this sad outcome is inevitable when one wants to glory in scientific progress without wanting to see the true glory of its origin. For science no longer radiates glory once cast in moulds strange to its uniqueness in history, moulds fashionable today but outmoded tomorrow.

THE PERSPECTIVE OF ORIGINS

Among the strange moulds into which science is being cast nowadays the mould of political revolution is the most fashionable but also the most counterproductive. The inner logic of such revolutions, all well known for devouring their protagonists or at least for turning into a rebuttal of their vision of progress, did not fail to claim its rights in that train of thought in which science is conceived as a chain of revolutions. The force of that logic clearly made it impossible to evaluate the history of science as a progressive march. The possibility that science was not such a march was to many a shock which hit especially hard some rationalist strongholds steeped in the dogma of a progress pivoted in science.[1] For, as the logic of the mould of scientific revolutions progressively unfolded itself, the history of science could not have a much better appeal than that 'ladder of misery, in which the revolutions form the different steps', a ladder that represented national history in Chateaubriand's *Essai historique, politique et moral, sur les révolutions anciennes et modernes*, first published in 1797.[2]

In giving this shattering definition of national history Chateaubriand was motivated by what in his eyes was a dishonest game which the *philosophes* played with truth. Part of that dishonesty, as portrayed by Chateaubriand, was their dismal failure to live up to the moral standards proclaimed by them. Thus was the stage set for a revolution leading to the convulsion of the Terror, which forced Chateaubriand to take refuge in London where he wrote the *Essai*. Its main thrust went far beyond the *philosophes* and the Revolution, or rather collapse, for which they played the role of midwife. Chateaubriand believed he had succeeded in showing that the causes of the collapse of classical Greece merely re-appeared in the factors that led to the débâcle of the France of the *philosophes*. Since he was convinced that the same parallel existed between half a dozen other revolutions or convulsions, ancient and modern, he concluded that history was inextricably caught in the treadmill of cyclic re-petitions.

Such was a grim conclusion though Chateaubriand hastened to add that it could produce very good effects. The principal of them was that 'every man who is persuaded that there is nothing new in history, loses a relish for in-novation, a relish that has been one of the greatest scourges with which Europe has been afflicted'.[3] Whether man can really live without living out his longing for something new was a point which Chateaubriand should have but did not broach. Indeed, within a few years he himself became a famous innovator. As the author of *Le Génie du christianisme* Chateaubriand certainly looked for and found something new. He would have firmly disavowed the suggestion that the novelty of Christian religion as a fulfilment of man's sense of beauty was a mere replay of ancient times which in his *Essai* seemed to go far beyond the deluge into a past with no apparent beginning. The author of the *Essai* was still to discover Christianity and its perspective anchored in the origin of all in the beginning.

The *Essai* was certainly true to its advocacy of a never-ending cyclic treadmill inasmuch as it exuded that pessimism which makes one find fault with almost anyone. Political and ecclesiastical leaders, as well as institutions, were ex-coriated by Chateaubriand no less than were ancient and modern sages. In that dismal picture there was but one un-

tainted flash of light but its significance was in all likelihood
lost on most readers of the *Essai*. Possibly even Chateaubriand, never a student of the sciences, failed to sense how
different from the entire tone of his book was that brief
paragraph in which he spoke of Copernicus, Galileo, and of
Newton. After all Chateaubriand's main interest 'in the
restoration of the true system of the universe' by Copernicus, in the invention by Galileo of the telescope, and in the
'immortal Newton' who 'stole from God . . . the secret of
nature', concerned the incidental detail that they in fact
owed nothing to the Encyclopedists.[4]

This was true enough, but Chateaubriand, once an eager
student of the *Encyclopédie*, merely repeated its shibboleths
about Copernicus, Galileo, and Newton. Even more was
this true of the brief encomium which Chateaubriand
heaped on Bacon.[5] But close as Chateaubriand's phraseology
was in that respect to that of d'Alembert's *Discours préliminaire*, he did not follow its author in labelling the rise of
science as a revolution. As if by instinct Chateaubriand spoke
of a restoration, not of a revolution.[6] The reason for this
was most probably a vision of science as something permanently acquired, a vision which Chateaubriand learned
from d'Alembert who popularized the phrase 'révolution
dans les sciences', a phrase slowly in emergence during the
first half of the eighteenth century.[7] Yet even the semantics
could not escape the logic of revolution, a logic which, although d'Alembert tried to evade it, was drawn before long
by Diderot who became an advocate of a perennial chaos
operating in the universe and a prophet of an endless chain
of revolutions in the sciences.[8] By that endless chain Diderot
could only negate the prospect of a steady forward march, a
prospect so dear to Condorcet but a prospect illusory in a
perspective of endless revolutions. The perspective was not

of progress but of continual relapse into chaos, a relapse brought into clear evidence by the Terror. It prompted Chateaubriand's dejected remark that most of the philosophy of the Encyclopedists 'is already forgotten, and all that remains is the French revolution'.[9]

All these details about Chateaubriand illustrate in a nutshell the logic of a basic alternative in choosing one's perspective on origin, a perspective with direct bearing on one's evaluation of science. The alternative played out its dynamics during those hundred or so years within which Chateaubriand's life fell and imposed a choice between two perspectives of cosmic and human history. One was the perspective of a linear march (though not in the sense of a naïvely optimistic progress) from an absolute beginning toward an absolute consummation, a perspective turned into a historical force through Christianity. The other was the perspective of no beginning with no end, a perspective inevitably leading to the mesmerizing vision of eternal returns.

To avoid taking any of these alternatives was the great ambition of the Enlightenment, an ambition anchored in the perfection and stability which the newly risen science displayed in its structure and attainments. The ambition received a classic expression in Kant's preface to the second edition of the *Critique of Pure Reason*. There Kant spoke of revolution in science as a sudden illumination, an unexpected 'glimpse', a 'happy thought',[10] anticipating with his psychologist metaphysics the present-day parlance about intellectual mutations. But unlike the moderns, Kant saw in that sudden illumination the opening of a secure path for science, a path not subject to further revisions or reversals. In mathematics this happened, so Kant declared, when the Ionians raised the Egyptians' practical art of measuring to the scientific level of geometry.[11] Physics entered its secure and final path when, to recall Kant's famous words, Galileo let balls roll down an inclined plane.[12] Tellingly enough, once the origin of science, or rather the beginning of its secure path, had been made so fortuitously simplistic, its future simply began to vanish. In Kant's view there was as little prospect for further progress in physics as there was for metaphysics, which he claimed to have subjected to a Copernican revolution and thereby put on a secure path.[13] To the lasting discomfiture of his admirers he did not spend his last years on working out the final details of metaphysics, a task of a few years in his estimate.[14] He rather spent those years on the codification of the final form of physics, a physics reduced to the five subjective senses,[15] in telling illustration of the subjectivism and un-Copernicanism of his metaphysics. It was part of the logic of all this that the old Kant, who had turned to pantheism, was all too happy with the sudden re-editions of his *Allgemeine Naturgeschichte* in which at the age of thirty he had described the universe as created in the form of homogeneous, infinitely extended matter but afterwards subject to an endless sequence of birth, death, and rebirth. Such a universe could easily do without a Creator, a sequel left to be spelled out with brute force by Kant's successors, right and left. They saw well enough that the middle-road dream of the Enlightenment about a cosmic and historical progress without origin and without end was a dream indeed. With the reality of God having been discredited the universe could readily become the ultimate reality, and as a reality in motion its motion had to be cyclic.

About the other alternative, a perspective of history, cosmic and human, anchored in the creation of all in the beginning and in the progress, though mysterious, of all te a heavenly consummation, Giambattista Vico was its only uncompromising and articulate proponent during the hundred or so years under consideration with his *Scienza nuova*.[16] Vico was certainly not original in speaking of the revolution of ages, a phrase which was very much in vogue from the late seventeenth century on and which ultimately inspired the phrase, revolution in the sciences.[17] He was certainly unusual with his spelling out the true spectre underlying the notion of laws of history operating in cycles or revolutions. In the measure in which Christian convictions were decreasing there increased the number of those who held less and less covertly that in accordance with eternal laws 'the affairs of all nations proceed in their rise, progress, mature state, decline and fall, and would do so even . . . if there were infinite worlds being born from time to time throughout eternity'.[18] But where the quotation is broken there was inserted Vico's parenthetical though emphatic disclaimer: '(as is certainly not the case)'. Vico, a devout Christian, could not advocate the pagan notion of Great Year and the equally pagan notion of worlds and cultures rising and decaying in interminable sequence. To interpret his thought about the succession of those cycles, which he saw completely subordinate to divine providence, one should rather think of a conical helix, the turns of which are drawn upward toward their apex.[19]

A conical helix, standing here for cycles in a history directed toward a supreme goal, can be such as to hardly differ, when not viewed from a proper distance, from a cyclindrical helix whose turns may symbolize the apparent progress of needless repetitions. In the same way Vico's rejection of the idea of endless returns through infinite time will reveal its significance only when seen in a context reaching far beyond its own times. The context is in a sense as far-reaching as man's cultural history. Until very recently it consisted in the history of a relatively few great cultures, not wholly independent of one another, yet sufficiently distinct in character. If there is a major characteristic common to them all it is their being steeped in a concept of cosmos and history subject to endless, repetitious cycles. That in all those cultures—Chinese, Hindu, Maya, Egyptian, Babylonian, to mention only the most significant ones—science suffered a stillbirth, can be traced to that mesmerizing impact which the notion of eternal returns exercised on them.[20] It was a mesmerism fomented by the lack of a firm foothold, which in turn could only be provided by the perspective of absolute origin, an origin inconceivable without belief in a creation out of nothing, an act implying a Creator.

That instead of the notion of a creation out of nothing only that of a generation out of nothing appeared on the horizon of ancient Greek philosophers[21] shows something of their addiction to a view of cosmos and history with no beginning and no end. Such a cosmos and history were trapped in the treadmill of endless repetitions which found many startling portrayals in the works of ancient Greek authors. Possibly the most memorable of them is the description in Plato's *Statesman* of the rattling and trembling of the world machine, prior to its starting another of its great cycles or Great Years.[22] Obviously, the development of crafts and sciences had also to be subject to the same circularity. In fact, Aristotle remarked that whatever could be contributed by the crafts to the comfort of life had already been achieved several times in former ages.[23] The remark, implying a smug evaluation of the present, could hardly be a stimulus for scientific curiosity and creativity. The stagnation of Greek science from the late fourth century on should not therefore be the cause of undue surprise.[24] Greek science had already been on a downhill course for some time before the Greeks lost their independence to Rome. The comfort which Polybius, the last great Greek historian, found in his country's political demise did not convey hope but resignation if not unabashed leer: Rome, the conqueror of Greece, was not exempt from the universal cyclic course in which triumph could only be followed by defeat.[25]

The only exception to that wholesale yielding to the logic of eternal returns was a belief stubbornly maintaining itself within a small people, the Jews, who time and again were ready to exchange that belief with the worship of idols, a worship tied to the worship of a nature going through an endless birth and death cycle.[26] Once that belief had become concretized in the life of the hero of the Gospels, a new and even more powerful challenge was laid to a view of cosmos and history deprived of origin and end. While the Greeks found nothing revulsive in the idea that Socrates was going to drink the hemlock an infinite number of times in infinitely many successive aeons,[27] a Christian could not bring himself to seeing Christ betrayed by Judas time and again if redemption in him had any meaning at all. The writings of Origen and especially of Saint Augustine clearly show that it was belief in the Gospel truth of a once-and-for-all redemption in Christ that gave strength to Christians to jettison the doctrine of Great Year.[28] This is not to suggest that Christians could not be lured by the mirage of the supreme embodiment of the notion of Great Year, the starry realm for ever in rotation. Astrology was rampant even through that age of faith, the Middle Ages,[29] and one can only be baffled by Roger Bacon's prediction of the final triumph of Christianity on the basis that major planetary conjunctions signalled the successive ascendancy of major religions.[30] But when the age of faith was being

overcome by the renaissance of the age of lust, the relapse
to the pagan belief of eternal returns quickly followed. The
tone of Machiavelli's *Florentine History* was already subtly
pagan with its covert affirmation of an endless sequence of
empires,[31] each having, as Machiavelli declared in *The Prince*,
only the purpose of securing its transitory strength and advantage, that is, of satisfying its lust for power by any and
all means.[32]

In the cosmic perspective the paganism of Great Year saw
an unabashed return in the writings of Giordano Bruno,
advocate of a perennial cyclic transformation of all into all,
who tried to make of Copernicus an ally in his advocacy
of that paganism in which the universe was an ever-living
animal.[33] Bruno's was a rank abuse of Copernicus whom he
really did not understand.[34] Bruno, the pantheistic pagan,
had no use for Copernicus' vision of the world as the
Creator's artifact having a permanence and rationality which
justified its interpretation in the lasting truths of an exact
geometry, which Bruno detested. Revealingly enough, it
was not Bruno's vagaries but Copernicus' vision that became a crucial factor of the scientific development which is
not the repetition of upheavals but a majestic genuine progress.[35] The vision of Copernicus, hardly a progressive man
in the sense in which two centuries after him that word came
to be used, derived from his believing with other Christians
in a type of progress outlined in Saint Augustine's *City of
God*. The heavenly city of the Enlightenment[36] was a
limitation to the earth of an ideal of progress noticed first
with one's eyes fixed on heaven, but the tactic proved that
no significant idea can function outside its genuine matrix,
an outcome which Vico's reading of the past had in a sense
adumbrated.

Thus the context which sheds light on Vico's rejection

of eternal returns reaches into times that followed him. The matrix in question was the perspective of absolute origin and end, the denial of which became a cornerstone in Comte's philosophy.[37] The consequences for his interpretation of the history of culture and especially for the history of science were nothing less than disastrous. The effects of this cast indeed a long shadow on the historiography of science in France during the latter part of the nineteenth century and even beyond.[38] The process was similar in Germany, dominated by the Hegelian right and left. In neither was there logical room for the Creator, a consequence amounting to a power vacuum. Since it cannot be endured too long, a substitute for the Creator was looked for in the alleged absoluteness of spirit (on the right) and in the presumed eternity of matter (on the left). The spirit was too vague to inspire scholarly concern for concrete facts, scientific or historical, whereas the matter was too crude to encourage interest in those aperçus which transfer material concreteness to the level of science and history. Unfamiliarity with a median route could only hinder the vistas of historians of science in late nineteenth-century Germany where a rather crude materialism had considerable success in presenting itself as the true voice of science.[39]

For a heady glimpse of the logic of the power vacuum in question one must turn to the writings of Nietzsche and Blanqui, resolute advocates of the idea of eternal returns. Their hatred of the Gospel was as much an integral part of that logic as was their abuse of science.[40] That they abused science would today be denied only by some very partisan minds.[41] The conclusion would be far from unanimous that their abuse of science was a consequence of the rigorous consistency with which they implemented their denial of absolute origin. A lack of unanimity concerning such conclusion would hardly have a support in any ambivalence on the part of either Blanqui or Nietzsche concerning that origin.[42] But a steady look at the glare of their stark errors seems to be evocative of the brightness of a truth hardly to the liking of that relativism which is *the* mould of interpretation nowadays.

About that relativism, calmly tolerant of contradictions and ready to take meandering for progress, one can easily forget that it is the logical result of a belief in progress which banned the vistas of absolute origin. Concerning the interpretation of history in general and the history of science in particular, this relativism usually appears in the guise of rejecting historicism which, following its criticism by Popper,[43] is often taken in a rather restricted sense. According to that restriction historicism is a predictive interpretation of a history subject to and embodying absolute laws and truths. That many like Plato (who is Popper's chief whipping boy though his real intention seems to be the chastising of Saint Augustine and the main stream of Christian tradition by bearing down on Plato's absolute truths)[44] have looked at history in such a way is undoubtedly true. But historicism has other meanings as well,[45] and seems above all to be a reaction against rationalism, against its set of atemporal absolute truths, which could seem but to float in mid-air as rationalism refused to anchor them in an absolute origin. More down-to-earth seemed therefore the par excellence historicist position that all truths are relative, conditioned as they are by the relativity of any and all historical phases or situations. Leading historicist historians have indeed been emphatic in pointing out the relativization of all ideas and explanations within the framework of historicism, of even those ideas which, in the words of Dilthey, science and philosophy have left intact. Meinecke, Mannheim, Troeltsch and others would have also said with Dilthey that their whole life was devoted to overcoming the ensuing 'anarchy of opinions'.[46]

Historicist relativism was obviously resting on a support with a built-in schizophrenia. On the one hand, the historicists tried to save the uniqueness of history from being crushed in the moulds of a rationalism patterned on mechanistic or Newtonian science. Collingwood's efforts were particularly explicit in that respect.[47] Such was an emancipation of the understanding of history from the understanding demanded by science, an emancipation which cannot be commended enough. On the other hand, quite different ought to be one's reaction to the historicists' escape to the apparently placid but in their depths very treacherous waters of relativism. That escape is supported by the tacit belief that while absolute truths are being jettisoned with respect to history, science somehow safeguards man's instinctive need for truths transcending the relative.

Such a schizophrenic outlook was not to remain too long safe from a rude jolt. The jolt did not consist in the crude misrepresentation of Einstein's work on relativity by not a few philosophers of science who eagerly seized upon it as the supreme proof that everything was relative. Apart from the logical error—committed mostly by logical positivists[48] —of this misrepresentation, any careful look at special and especially at general relativity would reveal that Einstein as a theoretical physicist was the most absolutist of all of them. The jolt in question came not from science but from a historiography of science caught in the logic of a historicism celebrated by Dilthey. The process started with a Koyré inspired by Bachelard and resolved to discredit the major conclusions of Duhem, and reached its logical height in Kuhn's relativist paradigms which made it impossible to vindicate that very progress which science alone seemed to evidence.

The full unfolding through Kuhn's paradigms of the relativist logic of historicism consists not so much in his calling into doubt the continuity of science, an achievement which sent shock waves through rationalist circles and jolted many a historicist philosopher of science from his pleasantly relativist though schizophrenic slumber. It rather consists in Kuhn's question—a question phrased so as to invite a negative answer—whether it is useful to think of the universe as a consistent unity of all, if the cultivation of science can flourish through a sequence of disconnected paradigms.[49] Kuhn's negative answer to this question flouted the whole history of science, for he could hardly be unaware of the plain fact that from Oresme through Copernicus, Galileo, Newton, and Maxwell to Planck, Einstein, and even beyond, all great creators of science found most useful, nay indispensable, for their scientific creativity, the belief that the universe is fully ordered. It seems indeed that in the shallowness of the philosophy of historicism there is no room for the basic truth about science, namely, that all science is cosmology.

The history of science, with its several stillbirths and only one viable birth, clearly shows that the only cosmology, or view of the cosmos as a whole, that was capable of generating science was a view of which the principal disseminator was the Gospel itself. It was the Gospel that turned into a widely shared conviction the belief in the Father, maker of all things visible and invisible, who created all in the beginning and disposed everything in measure, number, and weight, that is, with a rigorous consistency and rationality. That belief in His creation is the ultimate perspective with respect to origin. Such is also the perspective which,

when ignored by the historian of science consistently and
with a fearless logic, makes him turn into disconnected bits
that scientific enterprise which is still largely believed, and
not without good reasons, to have been a consistent continuum since its genuine rise half a millennium ago.

While one's account of the rise and progress of science
can fragment the image of science through the application
of a fearless though mistaken logic, the same account can
turn into a worn-out cliché if the perspective of the Christian
origin of science is evaded for some extraneous motivation.
Some such motivation must have been at work when the
result is the absence of any reference to the Christian faith
of so many from Copernicus to Newton. They obviously
had much more to do with the rise of science in the West
than the anonymous alumni of the maritime school established by Prince Henry the Navigator. The connection of
Christianity with science would seem to call for much more
than for quoting a story from the diary of Vasco da Gama.
According to the quotation he tells us how on having
reached India he chose a convicted shipmate for the dangerous assignment of going first ashore. The only person there
to understand Portuguese was a Jew, who to the shipmate's
words 'We are looking for Christians and spices' replied
curtly 'Go to hell!'

This story is the only reference to Christianity in *Science
and Western Domination* by Kurt Mendelssohn,[50] well known
for his work in low-temperature physics but unknown,
prior to his retirement, for works relating to the history of
science.[51] In the same book the chapter, 'Divine Beauty of
Simplicity', which deals with Copernicus and Kepler, starts
with the phrase: 'Thus at the beginning of the sixteenth
century, European man not only knew that the earth was
round, but thanks to the ingenuity of the Jews in Portugal,

he was also able to find his way about it.'[52] That such a phrase is part of a tactic aimed at putting historic Christianity into a deep freeze is suggested by the fact that the same chapter is conspicuously barren of references to the encouragement which Copernicus, Kepler, Galileo, and many others derived from their Christian faith to see a divine simplicity in the apparently irregular flow of phenomena. It is of a piece with that tactic that the book is offered to the non-Christian East as an instruction that it is possible to acquire the spirit of science without assimilating a certain metaphysical outlook characteristic of those centuries of Western history which witnessed the rise of science.[53]

The spirit in question means domination as well, of mankind no less than of nature. That it need not necessarily remain a tool of global domination by the West is undoubtedly true, but—regardless of the future position and strength of the West—science will remain a tool of domination in anyone's hands. The question therefore remains open about the proper and future use of science as a tool. For a satisfactory answer to this question the perspective of origin is as indispensable and enlightening as for questions connected with the past of science. That future seems rather beclouded. Gone are the days when scientists in their laboratories could freely indulge in any research their fancy prompted them to do.[54] Heated debates about the wisdom of disposing of nuclear waste in the ocean, city ordinances issued against departments of biology engaged in changing genetic structures, agonizing reappraisals about the threat posed to the air by those very engines that put society on wheels and gave it wings, and, last but not least, the hapless syndrome of man's wasting most of his resources on weapons which are the more destructive the more sophisticated science they embody—such are some of the signs that show science to be a very mixed blessing. Of course, science being merely a tool produced by man cannot as such be the culprit for the threats it poses. The issue at stake is the way in which the tool, science, is used by man.

It is now found to the surprise of many that the very same science, which was supposed since Condorcet's time to secure rationality in man, is energizing his irrational instincts to a degree unparallelled by any other factor. A surprising find to be sure but only because wide acceptance has been given to Condorcet's gospel that the more instructed a man is the more he is bent on rational action. It has indeed become a generally accepted belief that Condorcet and others were perfectly right in opposing the Gospel truth that there is something very wrong with man ever since his origin. It is now becoming all too evident, and through science, that the Gospel's perspective on man's origin, or rather on his being burdened with sin since his origin, is not so foolish a proposition as modern man has been conditioned to believe through a so-called scientific education.

Original sin as a perspective of origin is inseparable from the perspective of the absolute origin of all in the beginning. A proof of their inseparability is that today they are vindicated together by that very science which since Condorcet was supposed to discredit both. Just as the past of science cannot remain a meaningful continuum within a historicist interpretation disdainful of the idea of an absolute origin, the science of the future will loom threateningly ominous if man has to wear a blindfold about a sinful tragedy marking his origin. Without taking sin for what it is and without applying the cure of grace to it, all efforts to cope with man's ability to sin will end in further incitement to sin.

The inner logic of this fearful predicament will not, of

course, appear immediately in its true nature. The ever-fresh popularity of panaceas will keep distracting from the seriousness of the diagnosis and thus a problem already grave in itself will be further aggravated. Last-minute but lame admissions of being on a fatefully wrong course do as little good to anyone as to Captain Ahab. Reflecting on his being in the grip of his futile chase of Moby Dick, that big white whale, he could merely mutter, 'All my means are sane, my motives and objects are mad',[55] and go on with his maddening endeavour. In an age of hydrogen bombs, multiple warheads, cruise missiles, laser guns, and anti-satellite satellites, it is not even possible to say that all means are sane. Some means, however sophisticated scientifically, are intrinsically insane. The insanity is not, of course, an intrinsic property of the means and tools; it is rather the result of a chronic blindness of modern man. Wrapped in admiration of his scientific tools, he is unable to see that, although tools are always and at times spectacularly tangible whereas purpose is always intangible, this difference does not invalidate purpose itself.

It should be of much food for thought that science, which is a most purposeful endeavour aimed at the understanding and control of the physical, has been unable to foster a vivid appreciation in man for an understanding of purpose, let alone of that purpose of his which is spiritual. As if by fate, this inability of science was noted at the same moment when science had reached full maturity. The moment was the publication of Newton's *Principia* in 1687. In that year Leibniz remarked that mechanistic science was a support for atheists because the way in which it was cultivated left no room for consideration of purpose.[56] Rarely was a well-intentioned remark more mistaken. Revealingly, Leibniz's principal error was a point which concerned the perspective of ultimate origin, or God. In making that error Leibniz was a theologian misled by the apparent perfection of geometry as he declared that the divine mind, a mind certainly purposeful, was a mind intent only on geometry. From this followed the equally mistaken, and equally reductionist, inference that geometry was the area where purpose was to be found and justified. In concrete terms this meant that since mechanistic science was an embodiment of geometry, a mechanistic science including questions and evidences of purpose was in principle possible. One only needed, Leibniz argued, to put in science more geometry embodying the minimum and maximum principle, that is, laws such as the law of refraction, which revealed purpose inasmuch as it showed that light always acts purposefully by taking the shortest route.[57]

Leibniz was not yet through with the errors generated by his mistaken starting point. To his theological, philosophical, and scientific reflections he wanted to give a historical backdrop. He found it in the *Phaedo* where Socrates pleads for a new physics. Indeed, Leibniz was so happy with his find as to plan a full translation of *Phaedo* to confound the materialists of his time.[58] To be sure, the story of Socrates being ready to die in order to secure happiness for his soul through eternity has never been a story to the liking of materialists. While materialists would agree on the need for acting according to one's conscience and perhaps some of them would even be ready to die for their consciences, no materialist can agree with a Socrates who preferred not to let his soul live for eternity with the blot of sin on it. For Socrates the sin would have been giving by his escape from prison the appearance that he considered himself guilty of the charges brought against him. Socrates' friends, who arranged for his escape but found him adamant, went on to prove to him with arguments taken from Anaxagoras' mechanistic physics that with the death of the body the soul too met its end. To rebut his friends Socrates chose to argue, one of the most fateful arguments in cultural history, that Anaxagoras' physics had to be entirely replaced by a new physics. In that new physics questions about, say, the position of a body with respect to other bodies were to be replaced by the question whether it was best for that body to be in the position in which it actually was. Once the idea of such a physics had been granted, it was a foregone conclusion that bodies, including the body of man, acted for a purpose and therefore had a soul.

It was a fateful argument, because it prompted Aristotle to work out in full that new physics which, as it turned out, 'was worthless and misleading from beginning to end'.[59] It was that physics which postponed the rise of physics by almost two thousand years and also spelt the death knell of Greek science. Nothing of this was noticed by Leibniz. He certainly did not face up to the question why Socrates' suggestion, espoused by Plato and Aristotle, had become a norm for the rest of Greek philosophical and scientific posterity.[60] Espousing that norm was actually part of that 'failure of nerve'[61] which had been the lot of the Greeks since that fourth century which witnessed the start of their exchanging vistas of rationality with crass magic.

Such a rush from one extreme to the other can only be explained by a serious crisis of which Socrates' predicament was the best illustration. As Socrates himself insisted, he had as a young man been an enthusiast of Anaxagoras' mechanistic physics. It seemed to explain everything, yet it failed to account for what proved be the most important facet of life, the perspective of purpose. With the wisdom of hindsight it is but child's play to realize that Socrates should have kept mechanistic physics and cultivated in addition a train of thought germane to the reality of purpose. In other words, he should have taken the stance that both physics and metaphysics are needed to cope with the full range of man's experience and predicament.

That Socrates and many other great Greek thinkers could not perceive this elementary solution may seem indeed a childish error. But in order to avoid that error they would have needed a conviction that mechanism and purpose are not irreconcilable notions and that in order to save one the other need not be jettisoned. Such a conviction—the history of philosophy from the Greeks to the present is a proof—is not at all easy to muster. It was mustered first in a broadly based manner during the Middle Ages. It was then, in an age steeped in the Gospel, that man was able to convince himself that since both means and purpose are the products of the Creator's wisdom, they cannot be irreconcilable no matter how disparate and mutually exclusive they may appear. It was that conviction inspired by the Gospel that made possible that science in which data wholly devoid of purpose are investigated for a purpose that can only be recognized through a vision which man alone is able to achieve.

In the measure in which Western philosophy had abandoned Christian vistas, the conflict between means and purpose appeared to be more and more irreconcilable. To save both, Kant could only fall back on the distinction between pure and practical reason, a distinction for which his philosophy offered no justification.[62] No justification of purpose but only despair of it could come from the equivocation of T. H. Huxley and others who claimed the vindication of purpose through the mechanism of natural selection.[63] Despair about purpose has indeed set the tone of thought in recent times, a despair which is further aggravated by the destructive potentialities acquired by science. Such is a hapless situation matched by some hapless reflections on it. Clearly, there is some haplessness in that effort which now blames Christianity for the impending ecological disaster on the grounds that it was the Gospel that secured general consent to the words of Genesis: multiply and subdue the the earth.[64]

The argument would make some sense if it were accompanied by the recognition that the medieval state of mind nurtured by the Gospel has indeed been responsible for the rise of science. Responsibility for the effect, the misuses of science, implies responsibility for the cause. But the latter responsibility, which in this age of science appears to be the most coveted credit, the credit for the rise of science, is not attributed to Christianity when its mentality is blamed, for instance, by the noted historian of technology, L. White, for the ecological misuse of science.[65] It should not be surprising that an inconsistent stance on the part of a historian makes him also an inconsistent interpreter of history. For it is anything but consistent with the fact of Saint Francis' devoted support of institutionalized Christianity to represent him, as White does, as the embodiment of a purely charitable Christian spirit, wholly separate from the dominating spirit exuding from Christianity as an institution.

As is well known, Francis of Assisi was motivated by a dream, the meaning of which was a support to be given to institutionalized Christianity and not to a dream which is a Christianity devoid of institutional structure. Christianity, including its major institution, was a seamless garment for Francis who typically enough received praises only for his kindness and communing with nature in the programme outlined by Sarton for *Isis* and the movement it represents.[66] That garment is as much a single piece as is the Gospel

which, as the efforts of its rationalist critics have amply
demonstrated, one can have only an integral whole or one
will not have it for long. This is a point to be kept in mind
when one finds leading scoffers at Gospel, Christianity, and
God, suddenly realize that the problems of the scientific age
can be solved only with an unabashed recourse to Christian
love. Of course, the love in question cannot be mere senti-
mentality, but rather a love as strong and as hard as are the
problems which it is expected to resolve. Hard, in fact, were
the problems which Bertrand Russell faced in 1950 in his
lecture series at Columbia University. 'The Impact of Science
on Society', the topic on which Russell lectured, must have
appeared a painfully hard impact indeed for anyone aware
of the impending race from atomic to hydrogen bombs and
beyond. What made that infernal race appear even more
hellish was the hatred animating it. The racial hatred which
had almost ruined the world was followed by class hatred,
of which there is no end in sight yet. Trapped in that race
of destruction and hatred man could but feel, regardless of
his science, or rather partly because of it, completely at a
loss about the meaning of his existence. Since the recovery
of meaning could not be had without an escape from hatred,
Russell could assign as a remedy only a love which was
strong enough to cope with hatred. The only such love was
Christian love, the only love which ceases to be itself when
it ceases to be a love of one's enemies. To propose that love
as the only solution must not have come easily to Russell.
Forty years earlier he had made the philosophical and literary
scene with his panegyrics of the blind world of atoms with
no room for honesty, purpose, and love. Now he had to face
the cynicism of all those whom he had entertained for forty
years with his gospel of agnosticism and atheism: 'The root
of the matter is a very simple and old-fashioned thing, a

thing so simple that I am almost ashamed to mention it, for fear of the derisive smile with which wise cynics will greet my words. The thing I mean, please forgive me for mentioning it, is love, Christian love or compassion. If you feel this, you have a motive for existence, a guide in action, a reason for courage, an imperative necessity for intellectual honesty.'[67]

Such a testimonial to a love proclaimed only in the Gospel can but deeply gratify any disciple of it. Russell did not, of course, mean to endorse the Gospel. He merely found it desirable that the love proclaimed therein should prevail far and wide. Such a desire can, however, be realistic if it can be shown that a crucial part of the Gospel can be retained without the rest of it and especially without Him on whom it rests. While this is still to be shown, in spite of Comte and of many others' unconvincing efforts, history provides many an example that the love in question is a natural corollary once surrender to Christ has been achieved. That surrender requires in a sense no more and no less than what T. H. Huxley advocated with respect to facts. 'Sit down before fact as a little child, be prepared to give up every preconceived notion, follow humbly wherever and to whatever abysses nature leads, or you shall learn nothing.'[68] Long before Huxley held high the open-mindedness and docility of a child, the willingness to become like a child had been defined in the Gospel the condition of learning the most basic and most important truths. Surrender to Christ demands only the willingness to sit down as a child before the gigantic fact He is. Once this is done, in line with one's sincere belief in the scientific programme of sitting down as a child before fact and without any *a priori* discrimination between facts of nature and facts of history, to say nothing of an arbitrary discrimination among facts of history, one shall also find that Christ and his message of love can only be had in the perspective in which He gave himself and gave a love so strong as to cope even with hatred. The perspective is the perspective of twofold origin, the origin of all in the beginning and the origin of man's drama which, precisely because it is a drama, is spiritual. Within that perspective science too recovers its progressive unity and secures its hope of beneficial progress. Such a result should make more appealing further reflections on the origin of science and develop thereby the science of its origin.

NOTES TO LECTURE ONE

1. As can easily be gathered from the text of papers and discussions edited by A. C. Crombie under the title, *Scientific Change* (New York: Basic Books, 1963).

2. Ibid., p. 855.

3. Indeed, the second reference in the first page of Koyré's *Études galiléennes* (Paris: Hermann, 1939) is to Bachelard's *Le Nouvel Esprit scientifique* (1934) and *La Formation de l'esprit scientifique* (1938). Bachelard's psychological interpretation of the history of science will be discussed later.

4. Or, to quote Bacon's words, 'the great instauration of man's empire over the universe'. See *The Works of Francis Bacon*, edited by J. Spedding, R. L. Ellis, and D. D. Heath (1857–74; reprinted New York: Garrett Press, 1968), 3:523.

5. *Of the Advancement of Learning*, bk. 4, in *Works*, 4:372.

6. Letter to Dr. Playfere (1608) in *Works*, 10:301.

7. *New Organon*, preface, in *Works*, 4:40.

8. Of these three works no comment is needed on Descartes' *Discours sur la méthode*. The word 'prodromus' starts the full title of Kepler's *Mysterium cosmographicum* (1596) which is hardly ever quoted even in scholarly works as *Prodromus dissertationum cosmographicarum continens mysterium cosmographicum* . . . Undoubtedly, Galileo's *Nuntius sidereus* . . . is a title particularly difficult to render into English, but it seems to stand for more than a mere announcement about the realm of stars.

9. *Dialogue concerning the Two Chief World Systems*, translated by S. Drake (Berkeley: University of California Press, 1962), p. [5].

10. Ibid., p. 10.

11. First in his *Traité de l'équilibre des fluides* (Paris: chez David, l'aîné, 1744), p. xxxii, and then in his *Preliminary Discourse to the Encyclopedia of Diderot* (tr. R. N. Schwab; Indianapolis: Bobbs-Merrill, 1963), p. 79.

12. In Fontenelle's *Nouveaux Dialogues des morts* (see critical edition with introduction and notes by J. Dagen; Paris: Marcel Didier, 1971), published four times between 1683 and 1700, and half a dozen times during the next fifteen years, science is treated thematically but very briefly in both the first and the second parts, both of which consist of three dialogues: one among the ancient dead, another among the ancient and modern dead, and a third among the modern dead. In the first part Harvey is made to defend the superiority of modern medicine and physics against Erasistratus representing ancient science (pp. 186–92). In the second part Galileo is charged with the task of

showing to Apicius, the celebrated gourmet of classical times, that whereas new knowledge can always be found, this is not true of pleasures (pp. 318–25). Fontenelle's Harvey and Galileo show that Fontenelle was still to achieve a view in depth of what had taken place in science during the seventeenth century. There is a great distance from the Fontenelle of the *Dialogues* to the Fontenelle of the *éloges* of great scientists which he delivered in great number during the last thirty years as perpetual secretary of the *Académie des sciences*, a post which he relinquished in 1740 at the age of eighty-three. That science is not a topic either in the dialogue among the ancients or among the moderns shows something of the superficiality of Fontenelle's views on science during the closing decades of the seventeenth century. Needless to say, Fontenelle in the *Dialogues* neither raises the question of the origin of science, nor gives as much as a brief sketch of the history of science. That Fontenelle belongs to the eighteenth century transpires all too clearly in 'Bernard de Fontenelle: The Idea of Science in the French Enlightenment', by L. M. Marsak (*Transactions of the American Philosophical Society*, vol. 49 [1959], part 7).— Although the entire fourth and last volume of Charles Perrault's *Parallèle des anciens et des modernes en ce qui regarde les arts et les sciences* (1688–97; see facsimile reprint, Munich: Eidos Verlag, 1964) is devoted to the sciences, there is no effort made there to show why astronomy and physics had in recent times reached a higher degree of perfection. In astronomy Copernicus' view is treated as a hypothesis, whereas the section on physics is restricted to Descartes with no mention of Galileo's law of free fall, to say nothing of Newton's *Principia* already in print for ten years.

13. 'Areopagitica, or A Speech for the Liberty of Unlicenc'd Printing' (1644), in *Complete Poetry and Selected Prose of John Milton*, with an Introduction by C. Brooks (New York: The Modern Library, 1950), p. 714.

14. *The Vanity of Dogmatizing* (1661), p. 227. See *The Vanity of Dogmatizing: The Three 'Versions' by Joseph Glanvill*, with a critical introduction by S. Medcalf (Hove, Sussex: The Harvester Press, 1970).

15. The phrase, now best remembered as the motto of the Royal Society, is an abbreviated form of a passage of Horace, 'nullius addictus iurare in verba magistri' (*Epist.* I. 14). In that longer form, though with no reference to Horace, it was prominently quoted by William Barlow in his *Magneticall Aduertisments: or Divers Pertinent observations, and approued experiments concerning the nature and properties of Load-stone: Very pleasant for knowledge, and most needfull for practice, or trauelling, or framing of Instruments fit for Trauellers both by Sea and Land* (London: by Edward Griffin for Timothy Barlow, 1616), Preface, p. Bv. Barlow

was a disciple of Gilbert and his book might have been the source of popularity of Horace's phrase among members of the 'Invisible College' which owed at least as much to the influence of Gilbert as to that of Bacon.

16. On Boyle's unreserved admiration for Bacon and on efforts to lure him away from the Royal Society, see F. Jones, *Ancients and Moderns: A study of the Rise of the Scientific Movement in Seventeenth-Century England* (2d ed.; Gloucester, Mass.: Peter Smith, 1961), pp. 169–70.

17. *Experimental Philosophy, in Three Books: Containing Experiments Microscopical, Mercurial, Magnetical* (London: printed by T. Roycroft, for John Martin and James Allestry, 1664), p. 192.

18. *A Brief Account of the New Sect of Latitude-Men together with Some Reflections upon the New Philosophy* (London: printed and are to be sold in St. Pauls Churchyard . . . 1662), p. 23.

19. Ibid.

20. Ibid. Patrick also believed that 'learning and knowledge will breake forth like fire and pierce like Lightning through all impediments'. The brevity of Patrick's booklet (24 pp.) also betrayed its propagandistic character.

21. *The Advice of W. P. to Mr. Samuel Hartlib, for the Advancement of some Particular Parts of Learning* (London: Printed Anno Domini 1648), p. 2. In the first page of the twenty-six page long booklet Petty also declared that there was no need to give 'an exact definition or nice division of learning, or of the advancement thereof', because it has already been done 'so accurately by the great Lord Verulam'. Later, Petty spoke of Bacon's natural history as 'the most excellent' of its kind, 'an exact and judicious Catalogue' of particulars (p. 13).

22. *Mercury; or the secret and swift messenger shewing how a man may with privacy and speed communicate his thoughts to a friend at any distance* (London: printed by I. Norton, for John Maynard and Timothy Wilkins, 1641), p. 10. The context deals with the technique of coded and abbreviated writing and also contains the statement that what Bacon said of the topic in the *Advancement of Learning* presents in substance all that may be said on it.

23. For these and further examples of Boyle's encomiums on Bacon, see Jones, *Ancients and Moderns*, p. 170.

24. See 'A Proemial Essay' in *Certain Physiological Essays and Other Tracts* in *Robert Boyle, The Works*, edited by Thomas Birch (2d ed., 1772), reprinted with an Introduction by Douglas McKie (Hildesheim: Georg Olms Verlagsbuchhandlung, 1965), 1:305.

25. *Brief Lives chiefly of Contemporaries set down by John Aubrey between the Years 1669 and 1696*, edited by A. Clark (Oxford: Clarendon Press, 1898), 1:299. Henry Stubbe, one of the most vocal critics of Bacon

and of the Royal Society, also found it advantageous to discredit Bacon's memory with a reference to his ambitions: 'The only judgment I can make of my Lord Bacon's Actings, is, that being so Flagitious, and so ignominiously degraded: He determined to redeem the Infamy of his past Life by amusing the World with New Projects; and to gain a Chancellorship in Literature, when he was excluded from that on the Bench'. Stubbe also claimed that Bacon wanted to revenge himself on the nation when he wrote his books with the aim of 'creating in the Breasts of the English such a desire of Novelty' that could be but contemptuous of 'the Ancient and Civil Jurisdiction'. See the preface (p. 6) of Stubbe's *A Relation of the Strange Symptomes happening by the Bite of an Adder and the cure thereof: in a letter to a learned physician and a Reply by way of Preface to the Calumnies of Eccebolious Glanvil* (London: printed for Phil. Brigs, 1671). In the same year Stubbe also published two longer anti-Baconian booklets of which one was an examination of Bacon's views on sweating sickness. On commenting on Bacon's complaint in the *New Organon* (bk. 1, aph. 74) about the long decline of the sciences, Stubbe offered in rebuttal a brief history of science which deserves to be quoted in full because it shows that even around 1671 it was possible to see more in the scientific past than what the Baconians wanted their contemporaries to see: 'Who knows not the large improvements that the Mathematicks received by Euclid (who lived after Aristotle) and others, Who had then advanced every part of the pure and mixt Mathematicks long before the Verulamian philosopher writ this? Who knows not, how Herbary had been improved by Theophrastus, Dioscorides, the Arabians, and other Peripateticks, in like manner? Who can deny that Physick (in every part of it) and particularly Anatomy was improved by Erasistratus, Herophilus, Galen, Vesalius, Fallopius, etc before the Lord Bacon ever sucked? And what accessionalls had Chymistry recieved by the cultivation of the Aristotelians, before that the *House of Solomon* was dreamed of, or the *New Atlantis* discoevered?' (ibid.).

26. *New Organon*, Epistle Dedicatory, and bk. 1, aph. 122, in *Works*, 4:11 and 109.

27. Ibid., author's preface, in *Works*, 4:40.

28. Ibid., The Great Instauration, preface, in *Works*, 4:15. Clearly, the miscarriages or stillbirths in former times must have appeared uninstructive to him if he had no intention of recording them. Yet, the miscarriages should have appeared monumental in precisely that perspective, the domination of nature and of human society by the most inventive nation, in which Bacon saw science still to be born. The former times were those of great ancient cultures each of which would

have decisively changed the actual course of history, had science been born within its framework.

29. Ibid., *Works*, 4:14.

30. *New Organon*, bk. 1, aph. 79, in *Works*, 4:78.

31. Ibid., bk. 1, aph. 71, in *Works*, 4:72.

32. Ibid., bk. 1, aph. 73, in *Works*, 4:74.

33. See *De augmentis*, or the Latin translation of *On the Advancement of Learning*, bk. 2, ch. 4, in *Works*, 4:365.

34. *New Organon*, bk. 1, aph. 124, in *Works*, 4:110.

35. Ibid., bk. 2, aph. 5, in *Works*, 4:123, and *Descriptio globi intellectualis*, ch. 5, in *Works*, 5:511.

36. *New Organon*, Great Instauration, preface, and bk. 1, aph. 74, in *Works*, 4:14 and 75.

37. Ibid., Great Instauration, preface, in *Works*, 4:14.

38. In his *Turning Points in Western Technology: A Study of Technology, Science and History* (New York: Science History Publications, 1972), pp. 6–8, D. S. L. Cardwell emphatically notes in this connection the role of Christian faith according to which man is the God-appointed master of nature. Whenever the role of that faith is ignored (see for instance *Medieval Technology and Social Change* [London: Oxford University Press, 1962] by L. White, Jr.), the results leave some monumental facts of the historical record unexplained. According to White, the 'symptom of the emergence of a conscious and generalized lust for natural energy and its application to human purposes, is the enthusiastic adoption by thirteenth-century Europe of an idea which had originated in twelfth-century India—perpetual motion' (ibid., pp. 129–30). Yet the idea of perpetual motion failed to be productive in India. Indeed, as shown in my *Science and Creation: From Eternal Cycles to an Oscillating Universe* (Edinburgh: Scottish Academic Press; New York: Science History Publications, 1974; Chapter One), the failure of science to have a viable birth in India is due in a large part to the fact that Hindu cosmology never extricated itself from animism and that the universe was imagined there to be an animal going through, as if it were a perpetual motion machine, endless cycles of birth and rebirth. Whatever the debt of Latin medievals to 'the great Hindu astronomer and mathematician Bhaskara' (fl. 1150) who is credited by White (see his 'Tibet, India, and Malaya as Sources of Western Medieval Technology', *American Historical Review* 65 [1959], pp. 522–23) with the idea of perpetual motion machines, that idea had been known to them for some time in the Aristotelian notion of a heavenly mechanism forever in motion. Tellingly enough, the crucially important notion of impetus lies in that radical departure from Aristotle which can be seen in the commentaries of Buridan

and Oresme on Aristotle's *On the Heavens*. It was the Christian faith in an absolute beginning of the world that prompted both to assign to the Creator that original impetus which was given at the moment of creation to the entire created realm. See *Iohannis Buridani quaestiones super libris quattuor de caelo et mundo*, edited by E. A. Moody (Cambridge, Mass.: The Mediaeval Academy of America, 1942, pp. 180–81 (Lib. II, qu. 12), and Nicole Oresme, *Le Livre du ciel et du monde*, edited by A. D. Menut and A. J. Denomy, translated with an introduction by A. D. Menut (Madison: University of Wisconsin Press, 1968), p. 289 (Book II, chap. ii).

39. *Plus ultra: or, the Progress and Advancement of Knowledge since the Days of Aristotle. In an Account of some of the most Remarkable Endeavours. Occasioned by a Conference with one of the Notional Way* (London: printed for James Collins, 1668), p. 81.

40. Letter of December 23, 1630, to Mersenne, in *Œuvres de Descartes*, edited by C. Adam and P. Tannery (Paris: L. Cerf, 1897–1913), 1:195–96.

41. *Plus ultra*, p. 33.

42. *Discourse on the Method*, Part VI, in *The Philosophical Works of Descartes*, translated into English by E. S. Haldane and G. R. T. Ross (Cambridge: University Press, 1931), 1:119.

43. Quoted in *The Metalogicon of John of Salisbury: A Twelfth-Century Defense of the Verbal and Logical Arts of the Trivium*, translated with an introduction and notes by Daniel D. McGarry (Berkeley: University of California Press, 1955), p. 167. The context (Bk. III, ch. 4) should seem remarkable not only because it shows the medievals' resolve to see further than the ancients but also because it shows their awareness of seeing farther even than an Aristotle and in most decisive issues. Aristotle, John of Salisbury notes, failed to see about the notion of contingency certain facets which the Christian philosopher is able to see precisely because of his faith in the Creator.

44. Bk. I, pt. 2, in Migne, *Patrologia latina*, vol. 176, col. 205. Strangely enough, this incisive rendering by Hugo of St. Victor of the injunction given to man in Genesis is not quoted in R. W. Southern's presidential address, 'Aspects of the European Tradition of Historical Writing. 2. Hugo of St. Victor and the Idea of Historical Development', *Transactions of the Royal Historical Society* 21 (1971): 159–79, although he laid emphasis on the fact that part of the originality of Hugo's view of history relates to the significance accorded by him to the rise and cultivation of mechanical arts in the history of redemption. Yet, without keeping in mind Hugo's awareness of the importance of that biblical injunction, it is difficult to understand why those arts could be seen by him as a cure of the ignorance which had beset man following the Fall. Failure to recall that injunction makes it even

more difficult to understand the fact that this new conception of the meaning of mechanical arts acted as a driving force behind the scientific movement of the twelfth and thirteenth centuries, a fact noted by Southern. Although he gave no details of that movement, he recalled as its embodiment the words of true modernity of a late thirteenth-century author, Alexander Nequam, who extolled the cultivation of arts and sciences as the best source of power for the political state: 'The glory of a kingdom grows so long as the study of the liberal arts flourishes in it. For how shall enemies prevail over a kingdom which has mastered the sciences? The cunning skill of men, who have followed the secret flight of subtle essences into the bosom of nature, will not be overcome by the stratagems of their enemies' (ibid., p. 177). Roger Bacon saw, as is well known, the cultivation of sciences in this light, although in reference to strengthening Christendom. It is safe to assume that just as Nequam had merely echoed an idea already articulated by medievals preceding him, medieval writings, especially commentaries on the first chapters of Genesis, contain utterances similar to Hugo's incisive phrase. This should help put in deeper light Southern's remark that Hugo's handling of the meaning of history was much more illuminating than the 'stuff' he worked with. The 'stuff' was an undue reverence for biblical chronology and for a mostly mythological lore about the origin of various crafts and arts. That 'stuff' was to be proven a mere chaff in the eighteenth century which then was suddenly at a loss in sight of a vastly extended history, cosmic and human, for which anti-Christian rationalism admitted no origin and no end, let alone a twofold origin, a creation coupled with the fall of man, to be discussed in the last of these lectures. The crucial contribution of Christianity to the intellect as well as to the will of man was not the neatness of a short span of history with an almost visible origin and end. The contribution rather consisted in the broad instilling of the conviction that history *has* an *absolute* beginning and an end. Once such beginning and end were discarded from Western man's outlook on history, cosmic and human, even a fully grown science proved itself incapable of coping with the 'complexities of endless time, of immense cyclical movements and unimaginable depths of chance' that took the place of a clear historical picture which, as Southern noted, was 'an important aid to the development of scientific thought' (ibid., p. 160).

45. The frequent recourse of Sprat to the notion of 'genius of the time' or 'genius of the nation' was noted and illustrated by J. I. Cope and H. W. Jones in their edition with critical apparatus of Sprat's *History of the Royal Society* (St. Louis: Washington University Studies, 1958), p. xx.

46. As evidenced by the work of Buridan, Oresme, Crescas, and many others.

47. *History of the Royal Society*, p. 22.

48. The main topic of Jones' monograph quoted in note 16. The conflict received its classic satire in Swift's *A Full and True Account of the Battel Fought Last Friday between the Antient and the Modern Books in St. James's Library* (London: printed in the Year MDCCX).

49. *Plus ultra*, p. 65.

50. Ibid., p. 20.

51. Darwin's comment was prompted by his reading William Ogle's translation of *On the Parts of Animals*. See F. Darwin, *The Life amd Letters of Charles Darwin* (London: John Murray, 1888), 3:252.

52. *History of the Royal Society*, p. 14.

53. Ibid., p. 45.

54. *Plus ultra*, p. 54.

55. *Histoire du grand Royaume de la Chine situé aux Indes orientales* (Paris: Jeremie Petier, 1588). In crediting the Chinese with a 'most subtle mind' for their discovery of the art of printing, a 'most subtle matter', Mendoza failed to note the difference between the Chinese way of printing from page-size blocks and the Western technique of movable types (f. 82v–84r). Mendoza insisted that the Germans (Gutenberg) learned the art of printing from Chinese sources through land route (Moscow) and from the Arabs. In the next chapter (17) of book III of Part I of his work Mendoza listed the books brought by P. Herrade from China with special attention to their contents relating to the crafts and sciences (f. 84r–86v). Whatever of Mendoza's enthusiasm, the details given by him clearly showed the backwardness of the Chinese and their being captive to gross superstitions. An early English translation by R. Parke of Mendoza's work was published by the Hakluyt Society in two volumes in 1853–54.

56. *Nouveaux Advis du Grand Royaume de la Chine* (Paris: chez Rolin Thierry et Eustache Foucault, 1602). Longobardi claimed that all Chinese are literate 'with the exception of a small number of merchants, artisans, servants, and laborers' (f. 11v), but noted in the next page that unlike the Europeans, they are not fond of novelties. According to him the Chinese did no such bad things as 'killing, debauching, stealing and the like' (f. 14r). Longobardi also quoted Father Ricci's opinion that only the Europeans understood the contents of those old books (f. 9r).

57. *De christiana expeditione apud Sinas suscepta ab Societate Iesu* (Augsburg: apud Christoph. Mangium, 1615). Trigault noted the primitiveness of the Chinese way of printing books together with their great number and inferior quality. He concluded his chapter 'On Mechanical Arts

among the Chinese' with the words: 'I think their pride has its source in their ignorance of better things, and in the barbarous condition of their neighbors' (p. 22). Trigault's realization that the monosyllabic Chinese language gave rise to much equivocation was accompanied with his seeing in the ideographic Chinese script the basis for a universal language (pp. 26–27). The same inconsistency characterized his description of Chinese mathematics and astronomy as 'not mediocre' but also 'riddled with a thousand errors' (pp. 29–30). With no qualification he praised the excellence of Chinese moral philosophy (p. 29), a procedure that became a staple feature of subsequent China reports.

58. The original Italian was published in 1643. The French translation by Clovis Coulon was printed by S. and G. Cramoisy in Paris. Semedo not only saw the state of Chinese learning with the eyes of a Westerner, but also reported it in the Aristotelian classification of sciences 'to make more sense of it' (p. 68). He found the Chinese altogether ignorant of algebra but sufficiently instructed in geometry (pp. 73–4). He noted the great skill of Chinese physicians in making use of pulse-beats and commended them for not dictating to their patients how much to eat, 'because the body know it better' (p. 82). The total of ten pages (pp. 73–82) which Semedo allotted to his summary of all branches of learning among the Chinese could hardly contain much information even if Semedo had not treated his material in distinctly Western categories.

59. *Sinicae historiae decas prima res a gentis origine ad Christum natum in extrema Asia, sive Magno Imperio Sinarum gestas amplexa* (Munich: typis Lucae Straubii, 1658). The strictly chronological treatment was somewhat balanced by a lengthy *index rerum* (18 pp.). Revealingly, it contained only one entry on arithmetic and none on geometry.

60. The *Sinarum scientia politico-naturalis sive scientiae sinicae liber inter Confucii libros secundus* (24 ff.) contained the incisive remark that the Chinese had no science 'in our sense', and was published in the final instalment of the *Relations de divers voyages curieux qui n'ont point été publiés et qu'on a traduits ou tirés des originaux des voyageurs* IV Partie (Paris: chez André Cramoisy, 1672). Almost the second half of the *Sinarum scientia* was an account of Confucius' life.

61. *Nouveaux Advis*, f. 6v.

62. There was little justification for Needham for listing in the vast bibliography of Volume II of his *Science and Civilisation in China* (Cambridge: University Press, 1965), the works of Mendoza and Longobardi, if he felt that there was a 'compelling reason' for ending his historical introduction with the entry of the Jesuits in China (p. 148). The reason, as Needham specified it, was that from that moment on 'Chinese science fused with universal world science' (p. 149). Such a

reason could not, however, be compelling on the basis of Needham's view on the origin of science, to be discussed later, to say nothing of the fact that his historical introduction was full of references to the interaction of ancient Chinese science with all ancient cultures.

63. *Nouveaux Advis*, f. 15r.

NOTES TO LECTURE TWO

1. References will be to the first edition, Paris: chez Jean Anisson, 1696. In 1698 there followed the publication of a third volume, giving the whole history of the imperial edict in favour of Christian faith together with a 'clarification on the homage rendered by the Chinese to Confucius and to the dead'. That a second edition of this third volume was published in 1700 was a clear indication of the real point which Father Lecomte tried to make and which made his book the centre of immediate and heated controversy.

2. In Letter II in the second volume (the first volume contained eight Letters), which was devoted to 'the ancient and modern religion of the Chinese' (2:131–91), Lecomte stated that Fohi, the first emperor, kept seven special animals to be offered as sacrifices to 'the Sovereign Spirit of heaven and earth' (2:134), that Hoamti, the third emperor, built a temple to the Sovereign Master of heaven, and that it was 'most credible that the three subsequent imperial houses had kept the knowledge of [one] God for almost 2000 years' (2:136). As late as 800 B.C. idolatry had not yet taken hold of the people, so Lecomte claimed in the same context (2:146–47).

3. Thus Lecomte contrasted the Chinese preference for the traditional with the Europeans' lack of taste for anything 'except what is new' (1:150).

4. Ibid., 1:135.

5. Ibid., 1:442.

6. Ibid., 1:443. Lecomte's remark refers to the 'high point of perfection' to which geometry had been raised in Europe, a subject which, in his words, 'may pass for the masterpiece of human spirit' and which, together with the cultivation of logic, 'shall never penetrate the academies of the Chinese' (ibid.). This and the preceding quotation are from Letter VIII on the characteristics of the Chinese mind (1:439–508), in which there are many other comparisons between the

respective state of learning in Europe and in China.

7. See E. J. Aiton, *The Vortex Theory of Planetary Motions* (London: Macdonald, 1972), p. 114.

8. Ibid., p. 127.

9. References will be to the New American Library edition (New York, 1960) with a foreword by M. Cunliffe.

10. *Gulliver's Travels*, p. 187. Swift was not so lucky in specifying the distances of the two satellites from Mars and their periods of revolution, as he merely gave figures that satisfied Kepler's Third Law.

11. Ibid., pp. 201-2. The production was based on the haphazard occurrence of 'meaningful' sequences of words and anticipated the early twentieth-century lampooning of the same idea if one imagines the production of all works of Shakespeare by letting a large number of monkeys go on typing aimlessly for a sufficiently long time.

12. 18:8. Craig's booklet of thirty-six quarto pages was published in London (typis Johannis Derby) in 1699. A German translation by Johann Daniel Titius, professor at Wittenberg and better known for his formulation of the law of planetary distances, was published in Leipzig in 1775.

13. The project should be seen in the light of the claim made by many among the protagonists of the scientific movement that, as Sprat put it in his *History of the Royal Society* (p. 42), a scientist will endeavour to say 'so many things in an equal number of words'. The claim reflected the myopia of Baconian empiricists concerning the art of reasoning, scientific or otherwise, and became a bone of contention between the 'ancients and moderns'. For further details, see A. C. Howell, 'Res et Verba', in *English Literary History* 13 (June 1946): 131-42.

14. *Gulliver's Travels*, p. 203.

15. In the same context Hunter also noted: 'Too much attention cannot be paid to facts; yet too many facts crowd the memory without advantage, any further than that they lead us to establish principles.' Acquaintance with principles would then provide that knowledge of 'the causes of diseases' without which the surgeon 'is like an armed savage who attempts to get that by force which a civilized man would get by stratagem'. See John Hunter, *Lectures on the Principles of Surgery*, with notes by J. F. Palmer (Philadelphia: Haswell, Barrington, and Haswell, 1839), pp. 8-10. The text is that of the notes taken by H. Rumsey of Hunter's lectures in 1786-7.

16. Essay XVII was the second of the eleven new essays added in 1748 in the third edition of *Essays, Moral and Political* (Edinburgh: A. Millar) to the fifteen essays contained in the first (1741) and second (1742) editions. It appeared with considerable stylistic improvements in the

fourth edition (1753), an indication of Hume's fondness for it. Essay XVII was not only by far the longest among the twenty-six essays, but in view of the more than a dozen subsequent editions of the *Essays* during the remainder of the eighteenth century alone, it became also the most widely read but least discussed essay on the question of the origin of science and on its stillbirth in China in particular.

17. *Lettres édifiantes et curieuses écrites des missions étrangères par quelques missionnaires de la Compagnie de Jésus*, 34 vols. (Paris: 1702–76). For Parrenin's letters see 21:77–186, 24:1–91, and 26:1–85.

18. Parrenin's replies, especially the first one, contained copious extracts from de Mairan's letters.

19. *Lettres édifiantes*, 21:120, where Parrenin states that he had not found a single atheist among the Chinese!

20. Paris: P. G. Le Mercier. The first volume dealt with Chinese history, the second with arts and crafts, the third with religion, philosophy, ethics, and 'other sciences', the fourth with geography. It was in the section on 'other sciences' ('De la connoissance des Chinois dans les autres sciences', 3:264–89) that du Halde dealt with the achievements of the Chinese in logic, rhetoric, arithmetic, music, geometry, and astronomy, with the last taking up much of the section. According to du Halde, Europeans could not vie with the Chinese in the skill of computing large sums (p. 267), but he found their geometry 'rather superficial' (p. 268) and noted that other parts of mathematics were totally unknown to the Chinese. He described them as a very proud nation that cultivated the idea of its own greatness on the basis of ignoring other nations (ibid.). Du Halde also recalled the astonishment of the Chinese on being shown vacuum pumps, organs, clocks, carillons, thermometers, hygrometers, and small planetariums. In the section on Chinese astronomy (pp. 271–89) du Halde took the view that the interest of the Chinese in astronomical observations indicated that they were the descendants of the Hebrews who, according to him, colonized China shortly after the deluge (p. 271). Much of the rest of the section was taken up with summaries of the astronomical work of Jesuit missionaries in Peking who found the ancient Chinese observations of eclipses reliable and very helpful, but who also noted the subservience of Chinese astronomy to superstition. Copernicus was mentioned only in the oblique manner that, as in Europe, in China too experts are divided on the question whether the heavens turn around the earth, or the earth around the heavens, with the remark that, as in Europe, in China too, the second alternative is held only by a small number of people! (p. 273). Du Halde found the reasons in the backwardness of science in China in the absence of competition and recompense, a rather incongruous remark in view of his opening

question of the true nature of the subtlety of the Chinese mind: 'mais est-ce de cet esprit qui invente, qui pénètre, qui creuse, et qui approfondit? Ils ont fait des découvertes dans toutes les sciences; et ils n'en ont perfectionné aucune de celles qui nous nommons spéculatives et qui demandent de la subtilité et de la pénétration' (p. 264).

21. This famous call of Leibniz for Chinese missionaries is in his twenty-eight-page long introduction to excerpts from six Jesuit reports selected by him for his support of the Jesuit position concerning the value of Chinese religious tradition, published with no indication of publisher and place in 1697 under the title, *Novissima sinica historiam nostri temporis illustratura . . .* edente G. G. L. To justify his position Leibniz described Europe as a place where 'corruption had immensely progressed'. The description not only implied a curious oversight of some well-known corruptions in China, but was coupled with a baffling silence about European superiority in the sciences. 'Were we not superior', wrote Leibniz, 'to the Chinese in a superhuman matter, namely, in the divine gift of Christian religion, any judge instructed not about the shape of goddesses but the excellence of peoples would award the golden apple to the Chinese.' Leibniz's idealization of the 'practice' of the Chinese in matters concerning natural theology and ethics was typical of a speculative mind looking at facts from a distance which could easily make things appear in the best possible light, although the Jesuits had called, from the very start, attention to the widespread practice of stealing among the Chinese and to their inhumane abandoning of all unwanted and sickly newborn babies whose still living bodies were often eaten up by dogs!

22. See chapter on the Chinese; in *Œuvres complètes de Voltaire* (Paris: Garnier Frères, 1877–83), 11:176. The sinophile trait of Voltaire's thought has been the subject of numerous studies.

23. Ibid., pp. 186–8.

24. *Lettres de M. de Mairan au R. P. Parrenin, missionnaire de la Compagnie de Jésus. Contenant divers questions sur la Chine* (Paris: chez Desaint & Saillant, 1759). On the previous, private circulation of the letters, see p. iii. The three letters (pp. 1–153) were followed by a dissertation on ancient contacts between China and Egypt.

25. *An essay on the First Principles of Government, and on the Nature of Political, Civil, and Religious Liberty* (London: printed for J. Dodsley, 1768), p. 8.

26. 'Intégral [calcul]' in *Supplément à l'Encyclopédie* (Paris, 1776–77), 3:624; quoted in English translation in K. M. Baker, *Condorcet: From Natural Philosophy to Social Mathematics* (Chicago: University of Chicago Press, 1975), p. 8.

27. References will be to the modern English translation by J. Barraclough

with an introduction by S. Hampshire, *Sketch for a Historical Picture of the Progress of the Human Mind* (New York: Noonday Press, 1955).

28. See Baker, *Condorcet*, p. 344.

29. Quoted from a manuscript note by Baker, *Condorcet*, p. 345.

30. Paris: chez Ch. Ant. Jombert. According to the subtitle, it was a work in which 'one gives account of the progress [of mathematics] from its origin to our present days; where one sets forth the picture and development of the principal discoveries and the disputes to which they give rise, and the principal features of the lives of the most famous mathematicians'. Montucla's emphasis on 'discoveries' was an emphasis on 'success', which certainly limited his horizon, although it was much wider and deeper than the 'histories' of science on which he could rely. About Riccioli's *Almagestum novum* (1651) and *Geographia et hydrographia reformata* (1661) Montucla rightly remarked that these works, although rich in historical data, showed 'no sufficient appreciation of modern inventions' (Preface, vol. I, p. xix). He found that the *Tractatus algebrae historicus et practicus* of John Wallis (1684) could not be more inexact in matters historical (ibid.). Something more profound could be expected from G. D. Cassini, was Montucla's just comment (ibid., p. xx) about his essay, 'De l'origine et du progrès de l'astronomie' (1685). It was also true that J. F. Weidler's *Historia astronomiae* (1740) was a 'heap of passages merely stitched to one another, which are on occasion even contradictory; the whole being interspersed with citations of titles, pages, and editions, a true chaos . . . and not a guide sufficiently illuminated by the torch of criticism' (ibid., p. xix). Montucla could hardly benefit from Esteve's *Histoire générale et particulière de l'astronomie* (Paris: A.-C. Jombert, 1755), a work of three volumes, and written, as Montucla remarked, with 'légérité', but the mistakes pointed out by him in that work (ibid., p. xxiii) would have called for a more devastating appraisal. Goguet's and Fugère's history of ancient science (see note 59 below) was published while Montucla's work was printed. It was with a tongue in cheek that Montucla remarked (ibid., p. xxv) that he learned in that history only the obscurity of its topic.

31. *New Organon* (bk. 1, aph. 93). Montucla made no mention of the fact that Bacon also called for the recording of all reversals in scientific history, a point to be discussed in Lecture IV in connection with Whewell. Montucla's oversight in that connection was all the more curious as he referred to the outline in Bacon's *De augmentis* of the main topics to be discussed in the history of a science. Yet, just as the reversal of scientific progress was difficult to explain on Baconian principles, Montucla did not take the word 'revolution' for a reversal but for a great advance when he noted that 'the history of a science

is a work useful for and capable of contributing to its progress; to say nothing of the satisfaction which will be relished by every philosophical mind in contemplating its birth, growth, and revolutions' (Préface, vol. I, p. vii). At the end of that Préface Montucla extolled pure mathematics as the branch of science never subject to 'retrograde' motion, that is, a science in which 'error' could never obscure truth, whereas such was not the case with mixed mathematics, or physics. In his words, 'the march of pure mathematics was never disrupted by those shameful downfalls of which all other branches of knowledge show so many examples' (ibid., p. xxv). If such was the case, a study in depth of those downfalls should have formed all the more a part of the historiography of science, a part conspicuously missing in Montucla's work. It nevertheless represents so great a progress with respect to earlier works that its reading, as was aptly remarked recently, can still be profitable if done with caution. See article, 'Montucla, Jean Étienne', by K. Vogel in *Dictionary of Scientific Biography*, vol. IX (New York: Charles Scribner's Sons, 1974), p. 501.

32. In discussing the remotest origins of science Montucla was forced by his Baconian creed to speak of a great and steady progress of mechanical arts among ancient peoples (1:93). Fixation of Montucla's mind on steady progress, however illusory, in at least one branch of science, was responsible for his failure to see a major problem, or any problem at all, in the glaring upheavals evidenced in its other branches. Clearly, there was a problem to investigate, if the people of India, among whom, so Montucla claimed, astronomy was born, 'are at present immersed in the most profound ignorance concerning physical astronomy' (1:404). The problem was posed by that 'restlessness of spirit' which Montucla singled out as the most needed ingredient for the successful cultivation of the physical sciences, as he registered the fact that apart from their work in algebra 'the Arabs made no remarkable progress beyond the Greeks' (1:369). He failed to probe into the causes of that restlessness, just as he was not stirred to a restless investigation as he reported Friar Bacon's view of Robert Grosseteste as one who 'profoundly penetrated geometry' (1:428). Montucla rested satisfied with platitudinous phrases such that 'the thirteenth century was a time of enlightenment' in comparison with immediately preceding centuries (1:417), but still was not 'a century of genius', although he credited it with the 'most useful invention' of magnifying glasses (1:429). As to the fourteenth century—needless to say he knew not even the names of Buridan and Oresme—it was written off by him in two pages: its sole glory existed, according to Montucla, in the invention of the compass (1:438). It was an invention about which the Baconian in Montucla had to say that it became a reality through

'insensible [degrees] of growth', although the role of genius haunted him in this connection too. But all he said was that 'the fourteenth century presents to us, in certain respects, characteristics less brilliant than the thirteenth century, where one sees great enterprises and men who have shown sparks of superior genius' (1:438).

33. Available in facsimile edition (Paris: Albert Blanchard, 1968) with a foreword by M. Ch. Naux. The additional two volumes are largely devoted to the history of science in the eighteenth century.

34. Translated and edited with an introduction by R. L. Meek, *Turgot on Progress, Sociology and Economics* (Cambridge: University Press, 1973).

35. See ibid., p. 13.

36. 'On Universal History', ibid., p. 88.

37. Ibid., p. 103.

38. Ibid.

39. Ibid.

40. Ibid., p. 111.

41. Ibid., p. 88.

42. Ibid. In the same breath Turgot also opined that had Columbus died at the same age, 'America would perhaps not have been discovered until two centuries later'. It escaped Turgot, the empiricist or rather sensationist, that Columbus' and Newton's discoveries were very different in kind and that great literary accomplishments, mentioned by him in the same context, constituted still another kind of 'discovery'. Two hundred and fifty years later Einstein displayed a far greater insight into these matters when he pointed out that a great scientific discovery, such as his formulation of general relativity, would have, sooner or later, been made even if the actual discoverer did not exist, whereas a great artistic feat, such as Beethoven's C minor Symphony, would have never been written by someone else. Such was at least the gist of a remark of Einstein as reported by A. Moszkowski, *Einstein the Searcher. His Work Explained from Dialogues with Einstein*, translated by H. L. Brose (London: Methuen, 1921), p. 99.

43. 'On Universal History', p. 70.

44. Ibid., p. 71. Turgot could hardly envisage that a place would be named Concorde not under the impact of the teaching of the *philosophes*, but because of the catharsis following a Terror which came about not without some connection with their teaching.

45. Ibid., pp. 101–3.

46. *Sketch for a Historical Picture*, p. 163. The translation, 'All errors in politics and morals are based on philosophical errors and these in turn are connected with scientific errors', should be contrasted with the original which instead of 'scientific errors' concludes with 'des erreurs physiques', that is, obviously, 'errors in physics'. Yet Condorcet de-

cried as 'physicaille' the experimental investigations published in the *Mémoires* of the Académie des Sciences (see Baker, *Condorcet*, p. 42), in clear evidence of the short route from the extreme of sensationism to that of deductive apriorism.

47. *Sketch*, p. 17.

48. Ibid., p. 18.

49. Ibid.

50. Ibid., p. 169. The correlation was, however, mechanical to such extent as to include all intellectual labour of man.

51. Ibid., p. 172.

52. Ibid., p. 37.

53. Ibid., p. 95. Not even that much credit was given to the medievals by d'Alembert as he surveyed in his article, 'Experiméntal' (*Encyclopédie ou Dictionnaire raisonné des sciences, des arts, et des métiers*, vol. VI [Paris: Briasson, 1756], pp. 298–301), the history of experimental method which in his eyes corresponded to the history of science. Although he protested any interpretation of his survey as a slighting of studies of metaphysics and logic, he suggested in the same breath that their study be compressed in a few pages (p. 300). He distinguished two levels of experimental method: one of plain observation, the other of planned experimentation. The former he saw vigorously pursued among the Greeks of old, but the rise of the latter, that is, the rise of science, he entirely credited to Bacon's influence. Between the Greeks and Bacon there were 'the centuries of ignorance that lasted all too long' and 'the dark times' which deserved to be mentioned only because even then there existed a few 'superior geniuses who abandoned the vague and obscure method of philosophizing, left the words for things, and searched for more real knowledge in their own sagacity and in the study of nature' (p. 299). D'Alembert mentioned only two such 'superior geniuses', Gerbert and Friar Bacon, about whose work he offered no information other than that both were considered sorcerers because of their mechanical inventions, but while 'Gerbert became a pope, Bacon remained a monk and miserable' (ibid.). All this showed that d'Alembert was one of those, decried by him, for whom 'friar Bacon was very little known and just as little read today' (ibid.). Although d'Alembert read the works of Francis Bacon, he did not do so with the eyes of a historian but with the eyes of one blinded by the 'rebirth of a philosophy properly so called'. In that framework one could easily declare, as he did, that 'chancellor Bacon announced a large number of discoveries that have since then been made' (ibid.).

54. *Sketch*, p. 121.

55. Ibid., p. 115. A claim worthy of a book which started with the declaration: 'Man is born with the ability to receive sensations' (ibid.,

p. 3), a declaration whose chief and most fateful error consisted in its being a partial truth. It was his espousing sensationism that prevented Condorcet to assign any genuine creativity to the mind.

56. Ibid., p. 114.

57. Ibid., p. 38.

58. *Lettres sur l'origine des sciences, et sur celle des peuples de l'Asie, adressées à M. de Voltaire par M. Bailly, & précédées de quelques lettres de M. de Voltaire à l'auteur* (London: M. Elmesly; Paris: Frères Debure, 1777). Bailly continued his speculations in his *Lettres sur l'Atlantide de Platon et sur l'ancienne histoire de l'Asie* (same publishers, 1779). For a brief presentation of the reasoning and contents of both, see E. B. Smith, 'Jean-Sylvain Bailly. Astronomer, Mystic, Revolutionary 1736–1793', *Transactions of the American Philosophical Society*, Vol. 44, Part 4 (1954), pp. 465–7.

59. *Histoire de l'astronomie ancienne depuis son origine jusqu'à l'établissement de l'École d'Alexandrie* (Paris: Frères Debure, 1775), p. 18. The depth and incisiveness of this question is best appreciated if one takes for its background the three volumes of *L'Origine des loix, des arts et des sciences et de leur progrès chez les anciens peuples* (Paris: chez Desaint & Saillant, 1758) by A. Y. Goguet and A. C. Fugère. Already the division of the history of the subject matter into three parts—from the deluge until Jacob's death, from there until King David, and from there until the return from the Babylonian captivity—may reveal something of the uncritical character of the work for which the authors' infatuation with Baconian empiricism is to be blamed. Their recognition that during the previous hundred years there had been more progress in the sciences than ever before (1:262) would have been more than enough to make them search for the reasons of this and especially in a book on the origin of the sciences. Obviously, the difficulties experienced by the ancients in recording and transporting information could hardly appear a decisive factor in that respect. As Goguet and Fugère listed the happy conjectures of the Greeks about heliocentrism, tides, the flattening of the earth at the poles, and the habitability of the moon (a curious list to be sure), they could not be ignorant of the fact that those conjectures had been widely known among them (3:118–19). In noting that these happy conjectures had become a systematic science only in modern times, Goguet and Fugère had also a most momentous problem on hand but they failed to seize on it. Like many others, they placed the origin of science in the rise of the experimental method and did not care to ask why even the Greeks, in possession of so many happy conjectures, failed to advance to the level of systematic science. Goguet and Fugère credited almost entirely the Egyptians of old with the beginnings of science and gave but relatively few pages (3:315–46)

to ancient Chinese history. The ample leisure and peaceful times enjoyed by the Egyptians were, according to Goguet and Fugère, a main source of their progress, but they did not ask why, for instance, the Egyptians did not develop arched vaults, a point which they illustrated with various drawings of Egyptian monuments (3:70–1).

60. *Histoire de l'astronomie ancienne*, p. 18.

61. Ibid., p. 105.

62. A point amply documented in R. R. Palmer, *Catholics and Unbelievers in Eighteenth-Century France* (Princeton: Princeton University Press, 1939) and similar earlier studies listed there. This is why, as F. A. Manuel aptly noted (*The Prophets of Paris* [Cambridge: Harvard University Press, 1962], p. 71), those who after Condorcet wrote on the philosophy of cultural history could either accept or reject him.

63. As will be discussed in detail in Lecture Five.

64. In his letter of March 25, 1776, to Voltaire; see *Voltaire's Correspondence*, edited by T. Besterman (Geneva: Institut et Musée de Voltaire, 1953–65), 94:97. Bailly's primitive nation was seized upon by Antoine Court de Gébelin in support of his theory of an original universal language which he tried to reconstruct in a genuinely sensationist fashion from the thesis that soft sounds were invariably used to denote soft sensations and harsh sounds the harsh ones. No sooner had the first of the nine volumes of *Monde primitif, analysé et comparé avec le monde moderne, considéré dans l'histoire naturelle de la parole, ou grammaire universelle et comparative* (Paris: chez l'Auteur [etc.], 1774–82) appeared than the naïveté of sensationism had once more become evident through its concrete elaboration.

NOTES TO LECTURE THREE

1. Hegel could at best suspect that Kant was spending his last seven or so years on implementing an integral part of his philosophical programme, namely, the setting forth in broad outlines the form and division which physics was to have on the basis of epistemological postulates laid down in the *Critique*. The project was in a sense the 'moment of truth' of Kant's philosophy and even a brief look at Kant's legislation in physical science would explain the dismay and disbelief of Kant's admirers when from the 1880s on parts of his long-hidden manuscript began to be published. Complete publication did

not take place until the early twentieth century.

2. For a modern critical edition, see *August Comte. Écrits de jeunesse 1816–1828*. Textes établis et présentés par P. E. de Berrêdo Carneiro et P. Arnaud (Paris: Mouton, 1970), pp. 241–322.

3. In his letter of December 1824 to Valat; quoted ibid., p. 15.

4. See K. M. Baker, *Condorcet: From Natural Philosophy to Social Mathematics* (Chicago: University of Chicago Press, 1975), p. 40.

5. For details, see my *The Relevance of Physics* (Chicago: University of Chicago Press, 1966), p. 378.

6. 'From the beginning of my career, I have never ceased to respect the great Condorcet as my spiritual father', wrote Comte in his last major work, *Système de politique positive, ou traité de sociologie, instituant la religion de l'humanité* (Paris: chez l'auteur, 1851–4; 3:xv), which even by its title shows Comte's awareness of the fact that his smallish book published almost three decades earlier under the shorter title, *Système de politique positive*, contained the essentials of his positivist philosophy, a secularist theology in disguise.

7. From what Baker reveals in his *Condorcet* (p. 345) about the contents of the still unpublished 'Almanach', it is largely a list of the crimes and aberrations of a humanity lost in religious dreams.

8. Published in 1852, the *Catéchisme* stressed the 'positive' achievements of humanity. See modern edition with an introduction by P. Arnaud (Paris: Garnier-Flammarion, 1966).

9. Letter of April 22, 1851, to M. Tholouze, in *Correspondance inédite d'Auguste Comte* (Paris: Au Siège de la Société positiviste, 1903–4), 3:101.

10. For Huxley's phrase, see his 'The Scientific Aspects of Positivism' (1869), in *Lay Sermons, Addresses and Reviews* (5th ed.; London: Macmillan, 1874), p. 153. As was noted in the previous lecture, even Condorcet acknowledged the 'positive' contribution of religions already at the primitive stage toward a conceptual unification of various data of knowledge and gave particular credit to medieval scholastics.

11. *Catéchisme positiviste*, p. 265.

12. *Cours de philosophie positive. Tome cinquième contenant la partie historique de la philosophie sociale* (5th ed.; identical with the first; Paris: au Siège de la Société positiviste, 1894), p. 305. In the context Comte speaks of the social conditioning of the formulation of religious dogmas and that in such a perspective 'the intimately divine character attributed to the founder, real or ideal, of that great religious system' [Christianity] will appear a 'political necessity'. Contrasted with Saint Paul, Christ was, for Comte, merely a 'prétendu fondateur' of Christianity. See *Catéchisme positiviste*, p. 33.

13. *Essai sur les mœurs* in *Œuvres complètes de Voltaire* (Paris: Garnier Frères, 1877–85), 11:230.

14. Here again, as was the case with his formulation of the law of three states, Comte owed much to Saint-Simon. The latter, undoubtedly under the influence of Madame de Staël's *De la littérature considérée dans ses rapports avec les institutions sociales* (first published in 1801), insisted on seeing something positive in medieval institutions and faulted Condorcet's rejection of the Middle Ages as a stance incompatible with the notion of and belief in progress. (See G. Weill, *Un Précurseur du socialisme: Saint-Simon et son œuvre* [Paris: Perrin, 1894], pp. 61–3 and 133–4.) The restoration of the Middle Ages to cultural respectability became thematic in the works of such historians as Guizot and Michelet, but the respectability in question largely meant the fruitfulness of esthetical and social forces originating in Christian religion. When conscious effort was made to establish a positive connection between Christianity and science, the results were historically primitive and philosophically one-sided. This is well illustrated by the chapter on astronomy and mathematics in Chateaubriand's *Génie du christianisme*, first published in 1802 (see English translation, *The Genius of Christianity; or the Spirit and Beauty of the Christian Religion*, by Charles I. White [Baltimore: John Murphy, 1856], pp. 388–99). Of course, even if Chateaubriand had used the second edition of Montucla's *Histoire des mathématiques*, his account of medieval science would have remained primitive. As to philosophy, Chateaubriand could, of course, argue that one sound principle of morality was of greater value than all conclusions relating to quantities, but there was also an all-important *morale*, wholly overlooked by Chateaubriand, in the heuristic value of mathematics in the study of nature. The restoration of the Middle Ages to historical respectability made, however, more and more scholars turn to medieval manuscripts, but the choice of material showed the still prevailing conviction that the roots of modern science were anything but medieval.

15. The relevant documents are available in *Correspondance générale d'Auguste Comte*, ed. P. E. de Berrêdo Carneiro and P. Arnaud (The Hague: Mouton, 1973–6), 1:244–8 and 406–9, and 2:viii–xi. Comte had great hopes in Guizot because the latter had congratulated him in connection with the publication of his *Système de politique positive* (1825). Comte attributed the shelving of his petition by Guizot to 'sacerdotal influence'.

16. As the lectures progressed, so diminished the number of well-known scientists in the audience.

17. *Cours de philosophie positive*, 2:22.

18. He did so not only in the *Cours de philosophie positive* and in the

Système de politique positive, but also in the *Catéchisme positiviste* where he twice insisted on that restriction (pp. 98 and 112).

19. *Système de politique positive*, 1:511. The discovery could, according to Comte, be of interest only to the inhabitants of Uranus, the planet nearest to Neptune.

20. *Cours de philosophie positive*, 2:273–4. Apart from Kepler's and Newton's laws and from Laplace's studies on the stability of the solar system, the manifold aspects of astronomical studies appeared to Comte rather useless. He offered definitive considerations against ever measuring the parallax of stars, poured ridicule on efforts and hopes to investigate the chemical constitution of stars and felt that studies of the orbits of comets were reasonable only inasmuch as they prevented the recurrence of age-old superstitions.

21. He saw something antiscientific in efforts aimed at explaining chemical phenomena on the basis of electricity, attacked organic chemistry as having no value at all and urged its destruction, and called on biologists to resist the efforts of chemists to explain the phenomena of nutrition. Behind these and similar extravagances (for further details see my *The Relevance of Physics*, pp. 474–5) there lay an almost primitively sensationist philosophy which Comte inherited from Hume, whose professed admirer he certainly was.

22. Leçon XLV of the *Cours de philosophie positive*, which deals with the 'positive study of intellectual and moral, or cerebral, functions', is replete with glowing references to Gall, who also received special endorsements in the *Catéchisme positiviste* (pp. 32 and 299).

23. Although Comte felt that it was impossible to specify the beginnings of the 'great revolution' which ushered in the positivist age, he saw the 'combined impact of Bacon's precepts, of Descartes' concepts, and of Galileo's discoveries as the moment when the spirit of positivist philosophy began to assert itself in clear opposition to the theological and metaphysical mentality'. *Cours de philosophie positive*, 1:20. Prior to these three only the Alexandrine School of astronomers and some Arabs found favour with Comte as precursors of positivist thought.

24. Ibid., 5:3–4.

25. Ibid., 5:4.

26. It reveals something of the deeply ingrained character of certain clichés about intellectual history that even when the basic shortcomings of Comte, the historian of science, are recognized, he is still given credit for his 'erudition' in the same field. See, for instance, G. Sarton, 'Auguste Comte, Historian of Science', *Osiris* 10 (1952): 328–57, and L. Laudan, 'Comte', *Dictionary of Scientific Biography*, vol. III (New York: Charles Scribner's Sons, 1971), pp. 374–80.

27. As noted by Sarton, 'Auguste Comte, Historian of Science,' p. 346.

28. *Hegel's Philosophy of Nature*, edited and translated with an introduction and explanatory notes by M. J. Petry (London: George Allen & Unwin, 1970).

29. See F. Copleston, *A History of Philosophy. Volume 7. Modern Philosophy. Part I, Fichte to Hegel* (Garden City, N.Y.: Doubleday, 1965), p. 242. Fichte, in turn, would have referred to the difference between the purposeful and non-purposeful factors. Or as he put it in his analysis of the present age (*Die Grundzüge des gegenwärtigen Zeitalters* (1804) in *Johann Gottlieb Fichte's sämmtliche Werke*, edited by J. H. Fichte, vol. VII [Berlin: von Veit, 1846]), in so far as the non-purposeful elements were at work, there arose the factuality of that historical development which could but confuse with its detours, reversals, and apparent irrationalities. These could only be ascertained a posteriori in their particularities. But the philosopher could specify *a priori* the main course of history in its going through five epochs, (1) the age of the unconditional rule of reason through instinct, or the age of innocence; (2) the age of the expression of rationality in authoritative systems, or the age of rising sin; (3) the age of licentious emancipation from authority, or the age of universal sinfulness; (4) the age of scientific rationality, or the onset of justification; (5) the age of artistic rationality, or the age of complete sanctification. Scientific rationality was not, however, tied for Fichte to empirical science. He held high the notion of an *a priori* physics based on *Naturphilosophie*. About the reliance of a philosopher-scientist on empirical facts Fichte would have said what he said of his philosopher-historian: 'He starts with a *world-plan*, which in its unity lends itself to clear comprehension and from which the chief epochs of mankind's life on earth can fully be derived and can clearly be seen in their origin and mutual interconnectedness' (p. 6). If this postulate was sound then it followed that 'the philosopher, in so far as he deals as a philosopher with history, proceeds along the a priori track of the world-plan which is obvious to him without any [reference to] history; and his use of history is in no way to demonstrate anything with it, because his theses have already been proved and independently of all history; rather this use of his of history is only explanatory illustrating as it does that factual life in history which is comprehensible even without history' (pp. 139-40). Within the perspective urged by Fichte and the idealists there could be no urgency for a study of the concrete history of science let alone of its concrete origin.

30. Typically enough, the work, *Das Leben Jesu* (1795), in which Hegel grafts this dialectic on the historical Jesus, is Hegel's first work that shows Kant's influence on him. For further details, see Copleston, *Fichte to Hegel*, pp. 198-200.

31. As would be revealed by a brief look at the table of contents of his *The Philosophy of History*.

32. *Studies in Hegelian Cosmology* (Cambridge: University Press, 1901), p. 250.

33. G. W. F. Hegel, *Erste Druckschriften* (ed. G. Lasson; Leipzig: Felix Meiner, 1928) contains a German translation facing the Latin original, *Dissertatio philosophica de orbitis planetarum* (Jena, 1801). See especially, pp. 398–401.

34. See R. Wolf, *Handbuch der Astronomie, ihrer Geschichte und Literatur* (Zürich: Druck und Verlag von F. Schulthess, 1892), 2:455.

35. One of its victims was Georg Simon Ohm, discoverer of a basic law concerning electric current, who was denied for a long time a suitable academic post owing to his non-Hegelian approach to scientific questions.

36. In a letter written to H. Starkenburg from London on January 25, 1894; see *Karl Marx and Frederick Engels, Selected Correspondence 1846–1895*, translated by D. Torr (New York: International Publishers, 1935), p. 517. Forty years later, J. D. Bernal, an enthusiastic disciple of Engels, claimed in a context relating to Marxism and the history of science that the origin of Newton's theory of gravitation was a concern for problems of ballistics and navigation! According to Bernal only Marxism could secure from collapse a science left to the mercy of the only other alternative, the chance emergence of geniuses (see 'Dialectical Materialism' [1934] in J. D. Bernal, *The Freedom of Necessity* [London: Routledge & Kegan Paul, 1949], pp. 379–80). Another forty years later Marxists could at best explain away the still incomparably higher incidence of scientific geniuses within unregimented societies, to say nothing about the stifling of many a genius within Marxist realms. Bernal, of course, was no longer alive when Professor Margaret Gowing showed ('Science and Politics: the Eighth J. D. Bernal Lecture', 17th May 1977 [Birkbeck College, London, 1977]) that the development of nuclear energy in Great Britain defied at every step the norms set forth in Bernal's voluminous studies on science and society.

37. Those recourses are so numerous that in the Index of *Dialectics of Nature* (tr. C. Dutt [New York: International Publishers, 1940], references to Hegel are given as *passim*, the only entry having that dubious distinction. Numerous are the references there to Clausius, Helmholtz, and Newton, but only because they are considered chief enemies of dialectical materialism.

38. *Selected Correspondence*, p. 517.

39. *Popular Scientific Lectures*, translated by T. J. McCormack (5th ed.; La Salle, Ill.: Open Court, 1943), pp. 66–88; for quotation, see p. 67.

40. Ibid., p. 87.

41. Ibid., p. 85.
42. Ibid., p. 86.
43. Ibid., p. 88.
44. He did his best to prevent the establishment in Salzburg of a Catholic University free of State interference, and he called, in a full-page newspaper advertisement, on all Catholics to leave the Church following the promulgation of the Syllabus of Pius X. For further details, see J. T. Blackmore, *Ernst Mach: His Life, Work, and Influence* (Berkeley: University of California Press, 1972), pp. 235 and 290–2.
45. Mach's words, 'I have already declared and emphatically so that I am not at all a philosopher, but only a scientist', in the preface of his *Erkenntnis und Irrtum: Skizzen zur Psychologie der Forschung* (originally published in 1905; see 5th ed.; Leipzig: Johann Ambrosius Barth, 1926, p. vii), clearly implied that since he would not consider his use of history in support of his philosophy of science as an enterprise in philosophy, he could but reject its being evaluated as writing history. That the word 'history' appears only in the subtitle in some of Mach's writings should seem to be a subtle indication of this, and not, as O. Blüh wanted us to believe ('Ernst Mach as an Historian of Physics', *Centaurus* 13 [1968–9]:69) that Mach thereby merely expressed his wish to transcend the limits of that historicism which finds in history the evidence of objective truths.
46. 'As treated according to the method claimed by Mach, the history of mechanics will appear infinitely interesting to the physicist who seeks in the past only the light suitable to illuminate the present. If the historian and the psychologist were to forget that such was indeed the aim which the author wanted to achieve, they would undoubtedly complain.' *Bulletin des sciences mathématiques* 27 (October 1903): 269. The long review (pp. 261–83) is a masterpiece of carefully couched though devastating criticisms, which have been invariably neglected not only in the countless encomiums on Mach's work but even in critical appraisals of it.
47. Very eye-opening in this respect are the references to Duhem in the later editions of Mach's *Science of Mechanics*. While Duhem saw in his pioneering investigations of medieval science a rebuttal of the hallowed shibboleth of 'dark middle ages'—a point which was also a rebuttal of a principal aspect of Mach's view of scientific history—Mach restricted his acknowledgement of Duhem's researches to some technicalities concerning the contributions of some of Galileo's and Leonardo da Vinci's forerunners whom he even refused to call medievals.
48. This is certainly implied in the actual form of Mach's statement that 'if all the individual facts—all the individual phenomena, knowledge of which we desire—were immediately accessible to us, science would

have never arisen', a statement made in his first major publication (1872), *History and Root of the Principle of the Conservation of Energy*, translated and annotated by Philip E. B. Jourdain (Chicago: Open Court, 1911). The translation was made from the second edition (1909) which contains a few additional notes.

49. See Mach's inaugural address of 1883 as rector of the University of Prague, 'On Transformation and Adaptation in Scientific Thought' in *Popular Scientific Lectures*, p. 218.

50. See *The Autobiography of Charles Darwin*, with original omissions restored, edited with appendix and notes by N. Barlow (New York: W. W. Norton, 1958), p. 108. To what extent Spencer disarmed the critical faculties of some of his contemporaries can be gathered from chapter two, 'The Vogue of Spencer', in *Social Darwinism in American Thought* by R. Hofstadter (rev. ed.; Boston: Beacon Press, 1955).

51. In *Recent Discussions in Science, Philosophy and Morals* (New York: D. Appleton, 1871), pp. 155–234.

52. Ibid., pp. 158 and 161.

53. Ibid., p. 165.

54. Ibid., p. 185.

55. Ibid., p. 187.

56. Ibid.

57. Ibid., p. 212.

58. The three-volume work had by 1854 gone through two editions, and a third was to appear in 1857, with new impressions following in 1858 and 1865.

59. See chapter xxx, 'The Theological Bearing of the Philosophy of Discovery', in his *On the Philosophy of Discovery: Chapters Historical and Critical* (London: John W. Parker, 1860), especially pp. 372–4.

60. *History of the Inductive Sciences*, 3rd edition, reprinted (London: Frank Cass, 1967), vol. I, pp. 117–18.

61. Ibid., p. 199.

62. Ibid., p. 14.

63. Ibid., pp. 58–9.

64. *On the Philosophy of Discovery*, p. 124.

65. *Advancement of Learning*, bk. 2, in *The Works of Francis Bacon*, ed. J. Spedding, R. L. Ellis, D. D. Heath (London: Longmans, 1857–74), 3:330.

NOTES TO LECTURE FOUR

1. A remark of Sydney Smith, quoted in G. Himmelfarb, *Darwin and the Darwinian Revolution* (New York: W. W. Norton, 1968), p. 42.

2. Classic illustrations of this are *The Pioneers of Science and the Development of Their Scientific Theories* by O. Lodge, first published in 1893 and *The Growth of Physical Science* by Sir James Jeans, first published in 1947.

3. A. C. Crombie (ed.), *Scientific Change* (New York: Basic Books, 1963), p. 809.

4. The first to appear in print was *La Vie et l'œuvre de Pierre Duhem* by E. Picard (Paris: Gauthiers-Villars, 1921), which is the text of the 'notice historique' read by Picard in the public session of the Académie des Sciences on December 12, 1921. A year later an enlarged version with illustrations was published under the same title by the same publisher. 'L'Histoire des sciences dans l'œuvre de P. Duhem', an essay by A. Darbon, professor at the Faculté des Lettres of the University of Bordeaux, in *L'Œuvre scientifique de Pierre Duhem* (Paris: A. Blanchard, 1928), pp. 499–548, contains no information about the history and circumstances of Duhem's writing his *Système du monde* or about the motivations that made him turn to the study of the history of medieval science, which is also true of the essay of Picard, who was a mathematical physicist. Much of the volume in which Darbon's essay saw print is taken up by O. Manville's essay, 'La Physique de Pierre Duhem' (pp. 1–435). Only some fragmentary though very valuable details about the background of the writing of the *Système du monde* are available in Duhem's biography written by his daughter, Hélène Pierre-Duhem, *Un Savant français: Pierre Duhem* (Paris: Plon, 1936). For all its brevity, very reliable and useful for the historian and philosopher of science is P. Humbert's *Pierre Duhem* (Paris: Bloud et Gay, n.d. [1932]) which deals with the man, the physicist, the philosopher, the historian, and the teacher. The article 'Duhem, Pierre' by D. G. Miller in *Dictionary of Scientific Biography*, Volume IV (New York: Charles Scribner's Sons, 1971), pp. 225–233, contains but half a page on Duhem, the historian of science. This is curious, to say the least, because in the same *Dictionary*, professedly reserved for *scientists*, such non-scientists as Sarton, Koyré, and others received lengthy and glowing accolades because of their contributions to the historiography of science. It is difficult to explain why the task of appraising Duhem, the historian of science, was not assigned in that *Dictionary* to a competent historian of science, appreciative of and thoroughly familiar with Duhem's pivotal contribution to the modern historiography of science.

5. The most important of these are *Ernst Mach: His Life, Work, and Influence* by John T. Blackmore (Berkeley: University of California Press, 1972) and an essay by Gerald Holton, 'Mach, Einstein, and the Search for Reality', reprinted in his *Thematic Origins of Scientific Thought* (Cambridge, Mass.: Harvard University Press, 1973), pp. 219–260.

6. As can be seen from the list of Duhem's publications consisting of 352 items in *L'Œuvre scientifique de Pierre Duhem*, pp. 437–64.

7. Although not accepted as a thesis, Duhem's *Le Potentiel thermodynamique et ses applications à la mécanique chimique et à la théorie des phénomènes électriques*, running to 248 pages, was immediately published by the prestigious firm, Hermann (Paris, 1886).

8. *Un Savant français*, pp. 26 and 53.

9. Berthelot's anticlericalism is not aired in *La Pensée de Marcelin Berthelot* by A. Ranc (Paris: Bordas, 1948) and in *Vie de Berthelot* by L. Velluz (Paris: Plon, 1964), works that contain no reference to Duhem. In *Marcelin Berthelot: A Study of a Scientist's Public Role* by R. Virtanen (Lincoln: The University, 1965) Berthelot's anticlericalism is presented as a saintly attitude of not returning evil with evil (p. 22). No wonder that Berthelot's treatment of Duhem is not mentioned, although the failure of Berthelot's maximum-energy theorem is described in detail with Duhem being brought in briefly as 'that formidable enemy of positivism' who 'took special delight in making fun of Berthelot's principle' (p. 6).

10. He never did so in an ostentatious manner, let alone in a manner offending others' convictions. That he was described a 'religious extremist' (see D. G. Miller, 'Ignored Intellect: Pierre Duhem', *Physics Today* 19 [December 1966]: 53) is as much a sign of malice or miscomprehension as his being labelled there 'antisemitic'.

11. One wonders what Berthelot's reaction was on seeing the incisive review by Duhem of his *Thermochimie* come to a conclusion with a phrase from the *Imitation of Christ*, a favourite reading of Duhem. See *Revue des questions scientifiques* 12 (1897): 392.

12. The length of the letter (*Un Savant français*, pp. 158–69), written on May 21, 1911, from Bordeaux to Père Bulliot in charge of the newly established Institut Catholique in Paris, tells something of the depth of Duhem's feelings. In the letter Duhem describes the stereotype presentation of the rise of science as a liberation from the shackles of Church and Christianity, and adds: 'Devant cet enseignement, il est temps que l'enseignement catholique se dresse, et qu'à la face de son adversaire, il jette ce mot: mensonge!' (p. 163). I shall deal in my forthcoming monograph on Duhem with the very recent efforts of R. N. D. Martin to minimize the heuristic role of Duhem's Catholic

faith in his historiographic work ('The Genesis of a Mediaeval Historian: Pierre Duhem and the Origin of Statics', *Annals of Science* 33 [1976]: 119–29) and to present him as sympathetic to modernism.

13. *Un Savant français*, p. 150.

14. Ibid., p. 105. Half a century later a world-famous historian of science dismissed, in my 'clerical' presence, Duhem as a 'priest' and asked: 'Is it not strange that most of those interested in medieval science are Catholics?' He was most surprised to hear from me that Duhem was not a priest, and that this could easily be ascertained from his biography written by his daughter. Even without consulting any source, the same historian should have found something strange in the logic of his statement, namely, the implied disinterest in medieval science on the part of *non*-Catholic historians of science, professedly interested only in and all facts relating to their subject matter.

15. 'To condemn [the system of Great Year] as a monstrous superstition and to throw it overboard, Christianity was needed.' *Le Système du monde: Histoire des doctrines cosmologiques de Platon à Copernic*, vol. II (1914; reprinted, Paris: Hermann, 1965), p. 390.

16. After having reviewed and analysed the theses with cosmological relevance, Duhem noted: 'By hitting with anathema these propositions, Étienne Tempier and his councilors declared that in order to comply with the teaching of the Church [and in order] not to impose limits on God's omnipotence, the "peripatetic" physics had to be rejected. In doing so, they have implicitly called for the creation of a new physics which the intellect of Christians would accept. We shall see the University of Paris, in the fourteenth century, trying to construct that new physics and laying through that attempt the foundations of modern science. It was born, one may say, on March 7, 1277, from a decree issued by Monsignor Étienne, bishop of Paris. A principal objective of the present work will be the justification of this assertion.' *Le Système du monde*, vol. VI (1954), p. 66.

17. *Science and the Modern World* (New York: Macmillan, 1926), p. 2.

18. *To Save the Phenomena: An Essay on the Idea of Physical Theory from Plato to Galileo*, translated by E. Doland and C. Maschler, with an introductory essay by S. L. Jaki (Chicago: University of Chicago Press, 1969), p. 117.

19. See the meticulously and revealingly documented chapter 18, 'Mach and Buddhism' in Blackmore, *Ernst Mach*. The fact of Mach's Buddhism is still to become part of broader academic consciousness, a fact carefully kept under cover by Mach's admirers, especially by members of the Vienna Circle, an outgrowth of the Verein Ernst Mach.

20. The issue is the gravitational paradox of an infinite homogeneous universe, which should have been taken up by Mach as in his *Science*

of Mechanics he stated that the Newtonian concept of inertial motion implies assumptions about the totality of things, or universe. He sought escape in the limitedness of astronomical observations, adding the tell-tale phrase: 'it is certainly fortunate for us, that we can, from time to time, turn aside our eyes from the overpowering unity of the All, and allow them to rest on individual details' (p. 288).

21. As treated in full detail by G. Holton, see note 5 above.

22. It is no accident that Duhem's most explicit statement on this point is in his famous essay, 'Physics of a Believer', originally published in 1905, and added as appendix to the second edition (1914) of his *La Théorie physique: Son objet et sa structure*, first published in 1906. See the English translation by Philip P. Wiener of the second edition, *The Aim and Structure of Physical Theory* (1954; New York: Atheneum, 1962), pp. 293–305, especially, p. 298.

23. The most startling of these is contained in a letter of August 10, 1909, of Duhem to Mach in which Duhem stated: 'Believe, I beg you, in my profound respect and permit me to call your disciple' (quoted by Blackmore, p. 197).

24. In registering in *The Science of Mechanics* his indebtedness for information provided by Duhem on the forerunners of Leonardo, Mach carefully avoided calling them medievals. The Middle Ages remained for him the embodiment of obscurantism.

25. An omission all the more glaring because in the long review (*Bulletin des sciences mathématiques* 27 [1903]: 261–83) two pages (270–1) are reserved to the defects of Mach's treatment of the historical connections of mechanics with mystical, theological, and animistic notions.

26. The evidence is contained in the Preface which Mach wrote in 1913 to his *The Principles of Physical Optics* and, together with the book, was published only in 1920.

27. Analysis of the concepts and methods of science would show, Duhem stated, that 'it is possible without incoherence and contradiction to pursue the acquisition of positive knowledge and to meditate at the same time on religious truths'. The study of the history of science would in turn show, Duhem continued, that 'during those very ages when men cared above all for the kingdom of God and of its justice, God accorded to them as a bonus the most profound and fertile thoughts concerning matters down here'. *Un Savant français*, p. 168.

28. See note 31 to Lecture Five.

29. Published between 1927 and 1948 (Baltimore: The William and Wilkins Company).

30. See vol. XI (1960), p. 160.

31. *A History of Science: Ancient Science through the Golden Age of Greece* (Cambridge, Mass.: Harvard University Press, 1960), 1:3–4.

32. See *The History of Science and the New Humanism* (Bloomington: Indiana University Press, 1972), p. 74. But he also insisted that he was not a 'medievalist' and charged medievalists for having distorted the proper outlook on intellectual history 'because of their insistence upon the least progressive elements' and because of 'their almost exclusive devotion to Western thought when the greatest achievements were accomplished by Easterners'. *An Introduction to the History of Science*, 1:14.

33. *A History of Science*, 2:527.

34. *The History of Science and the New Humanism*, pp. 59–110.

35. Ibid., p. 78.

36. Ibid.

37. Ibid., p. 100.

38. See *Isis* 2 (1914): 204.

39. *Clerks and Craftsmen in China and the West: Lectures and Addresses on the History of Science and Technology* (Cambridge: University Press, 1970).

40. During the question-answer period following his lecture on China as teacher and healer of the West, given at the University of Edinburgh in the main lecture room, David Hume Tower, May 12, 1976. It is that modernism which reveals itself sheer religious syncretism in Needham's portrayal of 'The Historian of Science as Ecumenical Man' in *Chinese Science: Exploration of an Ancient Tradition*, edited by S. Nakayama and N. Sivin (Cambridge; Mass.: The M.I.T. Press, 1973), pp. 1–8, in which Needham's professed belief in Trinity is subordinated to the wisdom of Tao and his espousal of the mystery of the Cross is predicated on the altruism achievable in Tantrism! Highly characteristic of the 'ecumenism' of that book is the absence of Buridan and Oresme in its vast Index. The sole reference to Duhem is purely incidental and most misleading. The contribution which Duhem made 'to the great advances in the sophistication of the history of science' was due not so much to his 'positivism' than to his deep Catholic faith which enabled him to shed a vast and positive light on science in the Middle Ages, a light about which Marxists and positivists prefer to be silent for much the same reason.

41. See the preface to his *The Grand Titration: Science and Society in East and West* (London: George Allen and Unwin, 1969).

42. I have in mind the bibliographies of the first and especially of the second volumes of his *Science and Civilisation in China* in which the factual rise of science in the West is systematically discussed and where the role of historic Christianity should not have been evaded.

43. The occasion was the election of Duhem as corresponding member of the Académie des Sciences; see *Un Savant français*, p. 148.

44. R. S. Westfall, 'Newton and the Fudge Factor', *Science* 179 (1973): 751.

45. The request was made by Frederick Suppe, see F. Suppe (ed.), *The Structure of Scientific Theories* (Urbana: University of Illinois Press, 1974), p. 499.

46. A declaration made by I. B. Cohen, ibid., pp. 308 and 317.

47. See M. Masterman, 'The Nature of a Paradigm', in *Criticism and the Growth of Knowledge*, ed. I. Lakatos and A. Musgrave (Cambridge University Press, 1970), pp. 59–61.

48. *Études galiléennes* (Paris: Hermann, 1939), p. I-7, note 1.

49. Ibid., p. 1–6, note 2. Koyré referred to two works of Bachelard, *Le Nouvel Esprit scientifique* (Paris, 1934) and *La Formation de l'esprit scientifique* (Paris, 1938).

50. See chapter vi 'L'obstacle substantialiste', in *La Formation de l'esprit scientifique*, which carries the revealing subtitle, not mentioned by Koyré, *Contribution à une psychanalyse de la connaissance objective*. It was published again in 1947 (Paris: J. Vrin).

51. Ibid., p. 251.

52. Ibid., p. 250.

53. *Études galiléennes*, p. I-5. Koyré was quick to credit in that connection Meyerson, Cassirer, and Brunschvicg as well, as if their work had been as significant as that of Duhem.

54. Ibid., pp. I-9-10.

55. As pointedly and approvingly noted by C. C. Gillispie in his article, 'Koyré, Alexandre', in *Dictionary of Scientific Biography*, vol. VII (New York: Charles Scribner and Sons, 1973), p. 488.

56. *Un Savant français*, p. 155.

57. *From the Closed World to the Infinite Universe* (1957: New York: Harper Torchbook, 1958), p. 54. Why did Koyré have 'to regret to say' that Bruno was 'not a very good philosopher' and that he was 'a very poor scientist' who 'did not understand mathematics' and that his conception of celestial motions was 'rather strange'? Strange indeed it was and to a shocking degree.

58. Typically enough, Bruno did so in his *La cena de le ceneri* (1584), a work in which he set himself up as the genuine interpreter of Copernicus. See my English translation with introduction and notes, *The Ash Wednesday Supper* (The Hague: Mouton, 1975), p. 165.

59. The most defective in this regard is *La Cosmologie de Giordano Bruno* by Paul-Henri Michel (Paris: Hermann, 1962).

60. *The Works of Archimedes*, edited in modern notation with introductory chapters by T. L. Heath (Cambridge: University Press, 1897). The focusing of Archimedes' attention on statics is particularly revealing in connection with problems, such as the equilibrium of planes and the floating of bodies in water, which were also problems for non-uniform motion.

61. 'Every reader of medieval Latin texts knows that few Bible verses are so often quoted and alluded to as the phrase from the Wisdom of Solomon, 11:21, "omnia in mensura, numero et pondere disposuisti",' as pointedly noted by E. R. Curtius in his magisterial monograph, *European Literature and the Latin Middle Ages*, translated from the German by W. R. Trask (London: Routledge and Kegan Paul, 1953), p. 504.

62. See W. A. Wallace, 'The Enigma of Domingo de Soto: *Uniformiter difformis* and Falling Bodies in Late Medieval Physics', *Isis* 59 (1968): 384–401, and his *Galileo's Early Notebooks: The Physical Questions* (Notre Dame, Ind.; University of Notre Dame Press, 1976). In fact, to the end of his life Galileo retained the Aristotelian notion of scientific method as set forth in the *Posterior Analytic*, a method as different from its hypothetico-deductive brand glorified by the operationists as from the Neoplatonist mathematical realism advocated by Koyré. For an illuminating discussion of the cavalier handling of experimental data by which Galileo tried to overcome the shortcomings of the Aristotelian method, see W. R. Shea, 'Galileo and the Justification of Experiments', in *Historical and Philosophical Dimensions of Logic, Methodology and Philosophy of Science*, edited by R. E. Butts and J. Hintikka (Dordrecht: D. Reidel, 1977), pp. 81–91.

63. *Dialogue concerning the Two Chief World-Systems—Ptolemaic & Copernican*, translated by Stillman Drake (Berkeley: University of California Press, 1962), p. 104.

64. References to hedonism are much too frequent in *The Scientific Intellectual: The Psychological and Sociological Origins of Modern Science* by L. S. Feuer (New York: Basic Books, 1963) to require special documentation. Any instance given by him concerning the role of hedonism should appear as arbitrary and contradictory even in its context as does, for instance, his claim about China (p. 253). Thus Feuer states, in an obvious effort to protect his thesis against the objection to which familiarity with the age-long Hindu cultivation of sexual libido can most naturally give rise, that love played a part only in *modern* Indian philosophy! Again, only the most superficial reader would fail to be taken aback by Feuer's claim that the campaign in modern China for using the Latin alphabet was part of a rediscovery of the joys of sex (p. 261). Unquestionably, since the origination of science was a human process, it will show sociological and psychological origins, nay even origins relating to hedonism and libido. But these 'origins' when made to play the explanatory role reserved for the epistemological and ontological origin of science, or of anything else, will trap one in a disreputable shuffling, selecting, and interpreting of endless arrays of facts, an outcome also illustrated by Feuer's *Einstein and the Generations*

of Science (New York: Basic Books, 1974), a sequel to his *The Scientific Intellectual*. Here too the title of the book is a subtle give-away of its lack of serious merit. Undoubtedly, the antagonism of the younger generation toward the older, to say nothing of the 'hedonism' of the younger, explains many a thing, but the 'generations' of modern science will remain enveloped in the sociologist's and psychologist's jargon as long as the fact of generation or origin will remain divested of its epistemological and ontological connotation.

65. This mould, already used by Mach, has found in recent times its most spirited advocate in S. E. Toulmin. See his 'The Evolutionary Development of Natural Science', *American Scientist* 55 (1967): 456–71, and his *Human Understanding* (Princeton University Press, 1972), pp. 319–56. That in such mould science loses above all its claim to rationality is clear from the very motto of Toulmin's book in which rationality is identified with man's willingness to change his concepts. The motto can remain free of contradictoriness only if the concept of rationality need not be changed, in at least that very substance of it which being a 'substance' is hardly admissible within an evolutionary perspective that can admit no real permanence, the very meaning of substance.

66. The words, 'never use the word higher or lower', were jotted by Darwin on a slip of paper which he kept in his copy of Chambers' *Vestiges of Creation* as a constant reminder to himself. See G. Himmelfarb, *Darwin and the Darwinian Revolution*, p. 220.

NOTES TO LECTURE FIVE

1. As revealed, for instance, in the spirited argumentation of I. Scheffler in his *Science and Subjectivity* (Indianapolis: Bobbs-Merrill Company, 1967).

2. References are to the English translation, *An Historical, Political, and Moral Essay on Revolutions, Ancient and Modern* (London: Henry Colburn, 1815). For quotation, see p. 10.

3. Ibid., p. 395.

4. Ibid., pp. 356–7.

5. 'This great author opened the true road of philosophy to those who followed him; and every one, guided by his genius, knew from that time where to take his station', ibid., p. 353. In Chateaubriand's *Génie du christianisme* (see its English translation, p. 391, quoted in Lecture

Notes to Lecture Five

Three, note 16) Bacon is the crown witness against a science which in itself leads only to atheism!
6. Ibid., p. 356. For d'Alembert's use of 'revolution in the sciences', see his *Preliminary Discourse to the Encyclopedia of Diderot*, translated by R. N. Schwab (Indianapolis: Bobbs-Merrill, 1963), pp. 80–3.
7. The meticulous search by I. B. Cohen (see his 'The Eighteenth-Century Origins of the Concept of Scientific Revolution', *Journal of the History of Ideas* 37 [1976]: 257–88) yielded only three instances of the use of the word revolution in connection with the sciences during much of the first half of the eighteenth century. All three instances come from Fontenelle's *éloges* and all three refer to the revolution which in all likelihood would go on in mathematics following the discovery of infinitesimal calculus. Curiously, in the first two cases (1706 and 1719) Fontenelle was prompted by a mere textbook, L'Hôpital's *Analyse des infiniment petits*, first published in 1696 and the first of its kind. It was only in 1727 that Newton and Leibniz too were mentioned in such a context by Fontenelle, who emphatically spoke of Bernoulli, L'Hôpital, and Varignon as 'all great geometers who advanced the topic with giant steps'. Three phrases in twenty years by a perpetual secretary of the Académie des Sciences, who in fifty years in that office made it a rather staid institution, such is hardly a revolution even in semantics. No less unrevolutionary should appear the fact that for another twenty years no one, not even Fontenelle, seems to have repeated the phrase 'révolution dans les sciences'. Newton's *Principia* was in 1747 described by Clairaut before the Académie des Sciences as 'the epoch of a great revolution in physics'. But the phrase began to catch on only after d'Alembert popularized it in his *Discours préliminaire* and declared a few years later with a revolutionary consciousness that the revolution in the sciences had become a reality only in the generation that followed Newton. See his article, 'Expérimental' in *Encyclopédie ou Dictionnaire raisonné des sciences, des arts et des métiers*, vol. VI (1756), p. 299.
8. See L. G. Crocker, *Diderot's Chaotic Order: Approach to a Synthesis* (Princeton University Press, 1974), chapter I, and Diderot's article 'Encyclopédie', in *Encyclopédie ou Dictionnaire raisonné*, vol. V, pp. 636–637.
9. *An Historical, Political and Moral Essay*, p. 357.
10. *Immanuel Kant's Critique of Pure Reason*, translated by N. K. Smith (1929; New York: St Martin's Press, 1965), pp. 19–20.
11. Ibid., p. 19.
12. Ibid., p. 20.
13. Ibid., pp. 21 and 25. In accordance with this the *Critique* came to a conclusion with a brief survey of the principal 'revolutions' that had

taken place previously in metaphysics. See p. 667.

14. Ibid., pp. 668–9.

15. The evidence, still to gain sufficient publicity even in philosophical circles, and even among Kant scholars, is in the *Opus postumum*. It comprises most of the manuscript notes (several thousand sheets) which Kant piled up during the last seven years of his life. Interestingly enough, Comte, who certainly could not be familiar with those notes, arrived at a similar codification of physics on a basis which though not Kantian, was in final analysis not less subjective.

16. It was seen through press by Vico in three successively improved versions in 1725, 1730, and 1744. This last edition was translated into English by T. G. Bergin and M. H. Fisch, *The New Science of Giambattista Vico* (Ithaca, N.Y.: Cornell University Press, 1948).

17. For a large number of examples of that vogue, see the article of Cohen quoted in note 7 above.

18. *The New Science of Giambattista Vico*, p. 372.

19. This is in substance the main point made in chapter iv, 'The New Science of Vico', of *The Sense of History: Secular and Sacred* by M. C. d'Arcy (London: Faber and Faber, 1959), see especially pp. 129–31. Such a point will, of course, be held irrelevant by all those interpreters of Vico who, like Sir Isaiah Berlin (see his *Vico and Herder: Two Studies in the History of Ideas* [New York: The Viking Press, 1976]), admit with him on the one hand that Vico was 'a devout Christian' (p. 30), who 'during the entire second part of his life—his most creative years —lived in the most intimate intercourse with priests and monks, and looked to them for sympathy, help, advice, protection' (p. 79), but on the other hand are forced to claim, as Sir Isaiah is owing to his commitment to a non-transcendental humanism, that Christianity (to speak nothing of Catholicism) was an extraneous ingredient in Vico's system (p. 80). But could Vico be truly the victim of a dichotomy of mind if it is also true that he 'insisted on a personal God, the transcendental deity of orthodox Catholic Christianity' (p. 81) and that he was 'a Christian teleologist no less than Augustine or Bossuet' (p. 66)? For even if he was a victim of that dichotomy, was he to that extent as to abolish in his philosophy the difference imposed by the very first article of his orthodox Catholic faith between two kinds of creating: one strictly reserved for God, the Creator, and another in which man amuses himself by 'creating' his own field of interest? For only if Vico had abolished in his philosophy that difference, and in however surreptitious a manner, can he be turned into a prophetic forerunner of that modern humanism in which man can create not only his field of interest but even himself by identifying with an evolutionary process radically severed from a transcendental Creator

and thereby trapped in the treadmill of eternal returns. True, the orthodox Catholicism of Vico's time offered not much help for an evolutionary view of human society. But Thomas Aquinas found no help either in the orthodoxy of his time as he espoused Aristotle, a fact which in Sir Isaiah's evaluation of Vico should make Aquinas appear if not a heresiarch, at least a schizophrenic. A failure to espouse the orthodox doctrine of creation will in the long run fail one in his reasoning, a truth which holds even of *Vico and Herder,* but not of a Vico who kept a faith in the Maker of all which was the unfailing foundation of his philosophy safe of the mirage of a progress tied to the denial of transcendence.

20. For details and documentation, see in my *Science and Creation: From Eternal Cycles to an Oscillating Universe* (Edinburgh: Scottish Academic Press, 1974), the first five chapters dealing respectively with those cultures.

21. Generation out of nothing, as it implied no Creator, could easily be rejected as an illogical notion by Aristotle and others. The endorsement—the only one of its kind by an ancient Greek philosopher—of creation out of nothing by Atticus of the Middle Academy in the second half of the second century (A.D.) could hardly be exempt of Christian influence. Whether Lucretius' rejection in his *De rerum natura* (1:150) of the arising of something out of nothing, even if this should happen *divinitus,* indicates a rebuttal on his part of some 'theists', is rather debatable. At any rate, none of them is known by name, and certainly not with a reputation of a philosopher. The three extant passages in which generation out of nothing is rejected by ancient Greek philosophers are presented with great care in 'Creatio ex Nihilo' (*Studia theologica* 4 [1951]:13–43; see especially, pp. 23–6) by A. Ehrhardt, who is rather suspicious, partly because of his epistemology or rather the lack of it, of the import and genuinely Christian character of the doctrine of creation out of nothing. Thus he has only a few lines in his monograph, *The Beginning: A Study in the Greek Philosophical Approach to the Concept of Creation from Anaximander to St John* (New York: Barnes and Noble, 1968, p. 167), for the famous passage in 2 Maccabees (7:28) in which he sees the influence of Greek ideas! On Atticus, see ibid., p. 166.

22. See *The Collected Dialogues of Plato Including the Letters,* edited by E. Hamilton and H. Cairns (New York: Pantheon Books, 1963), p. 1038.

23. It should seem most revealing that Aristotle made this statement in his *Politics* (1329b). The cyclic fortune of crafts and arts was for him a proof that one should expect political developments to be subjected to the same pattern. See *The Politics of Aristotle,* translated by B. Jowett

(Oxford: Clarendon Press, 1885), 1:223-4.

24. It should, however, be surprising to those who see science as a function of material conditions. Yet, as G. E. R. Lloyd notes in his *Greek Science after Aristotle* (New York: W. W. Norton, 1973), 'the continuity in the conditions under which scientific research was conducted before and after Aristotle is striking' (p. 6). Therefore other than material conditions should be responsible for the fact that 'no *new* rationale or justification for scientific research was developed after Aristotle' (p. 7).

25. *Polybius, The Histories*, with an English translation by W. R. Paton, vol. III (London: William Heinemann, 1923), p. 277. In the same context Polybius also notes that democracy is the ruination of a society which does not couple it with a high measure of self-discipline, a point also highly applicable to the democracy demanded by broadly based scientific research.

26. At one point, during Manasseh's more than fifty-year long reign, even the Book of the Law lay forgotten in an obscure corner of the Temple which itself was the scene of cultic celebrations adapted from Canaanite tribes.

27. It is, of course, doubtful that the focusing, for instance, by Origen (see my *Science and Creation*, pp. 173-4) on Socrates' heroism as pre-empted by its endless recurrence, had made much impression on the adept of any of the three major Greek philosophical persuasions, the Platonic-Aristotelian, the Epicurean, and the Stoic, each of which derived its main inspiration from Socrates' life.

28. For statements of Saint Augustine to that effect which represent the high point of his *The City of God*, see my *Science and Creation*, pp. 178-181.

29. See, for instance, T. O. Wedel, *The Mediaeval Attitude toward Astrology, particularly in England* (New Haven: Yale University Press, 1920).

30. See *The 'Opus Majus' of Roger Bacon*, edited with an introduction and analytical tables by John H. Bridges (Oxford: Clarendon Press, 1897), vol. I, pp. 253-8.

31. Underlying that sequence was the antagonism of two factors, war and peace, of which the latter was frowned upon by Machiavelli because it invited the study of letters, a most seductive means to corrupt 'the vigour of warlike minds' who alone could secure stability. See *The Florentine History*, translated by N. H. Thomson (London: Archibald Constable, 1906), 2:1.

32. In order to keep his state prosperous, the prince was advised to act, if necessary, 'against faith, against charity, against humanity, against religion', because, Machiavelli declared, 'the end justifies the means'. See *The Prince* translated by L. Ricci, revised by E. R. P. Vincent, with

an introduction by C. Gauss (New York: The New American Library, 1952), pp. 93-4.

33. By advocating such an idea of the universe Bruno opposed that all-important trend in physical science and astronomy which aimed at shifting scientific thinking from the concept of organism to that of the machine as a basic explanatory framework. See on this the first two chapters in my *The Relevance of Physics* (Chicago: University of Chicago Press, 1966).

34. Indeed, Bruno misconstrued Copernicus' thought to that extent in the *Cena de le ceneri* that its study prompted F. Yates to remark in her *Giordano Bruno and the Hermetic Tradition* (Chicago: University of Chicago Press, 1964, p. 297): 'Copernicus might have well bought up and destroyed all copies of the *Cena* had he been alive!'

35. Tellingly enough, Bruno did not appeal at all to the Encyclopedists loudly proclaiming the gospel of progress. For their forerunner, Pierre Bayle, Bruno was the epitome of obscurantism, whose 'principal doctrines are a thousand times more obscure than all the most incomprehensible dicta ever uttered by the followers of Thomas Aquinas or John Duns Scotus'. See article 'Brunus' in Bayle's *Dictionnaire historique et critique* (1697; Paris: Desoer, 1820), 4:176.

36. Historians of ideas who idolize the Enlightenment understandably resent the unfolding by Carl L. Becker (*The Heavenly City of the Eighteenth-Century Philosophers* [New Haven: Yale University Press, 1932]) of not only its pseudoreligious roots but also its collapse into a treadmill which Becker's concluding words evoke through a quotation from Marcus Aurelius! Becker may have, of course, been influenced by that sobering chapter in *The Idea of Progress: An Inquiry into Its Origin and Growth* (1932; New York: Dover, 1955) which brings that work to a close and in which its author, J. B. Bury, admits that the secularized idea of progress is insufficient to vindicate the absolute truth attributed to that idea.

37. Comte's declaration that 'all idea of *creation* properly so-called must be radically avoided', is part of his introduction to cosmogonical theories in his lectures on astronomy in the *Cours de philosophie positive* (Paris: Éditions Anthropos, 1968), 2:280. A most appropriate context for such a declaration, as of all branches of science it is cosmogony that intimates most persistently the fact of creation.

38. This positivist influence was so strong as to secure the appointment of Pierre Laffitte, not a historian of any kind but head of the Positivist Church, as the first occupant of the chair for the history of science established at the Collège de France in 1892. Laffitte's incompetence could only discredit the field he was supposed to promote, as was also the case when he was succeeded in 1903 by Grégoire Wyrouboff, again

not a historian, but who also had strong ties with Comtean religion and its power base in the academic establishment headed by M. Berthelot, whose *Les Origines de l'alchimie* (Paris: G. Steinheil, 1885) is a classic example of a highly acclaimed dabbling in the history of science. Berthelot and his allies could not reconcile themselves to a devout Roman Catholic such as Paul Tannery who had by far the best credentials for the chair. Moreover, Tannery always insisted on 'positivism', that is, a meticulous attention and study in depth of source material. While this 'positivism' served him well in editing (his editions of the complete works of Diophantes of Alexandria, of Fermat, and of Descartes are examples of scholarship) and in research of factual nature (well illustrated by the seventeen volumes of his *Mémoires scientifiques*), it incapacitated him for the work of interpretation which alone can unfold history latent in chains of events. Thus in his *Pour l'histoire de la science hellène* (Paris: F. Alcan, 1887), Tannery stated (p. 10) that 'whereas the history of philosophy ought to be completed by the history of science, the latter, far from resting on the former, must be worked out independently by a totally opposite method'. Consequently, one would look in vain in that work even for a hint that the rise and decline of Hellenic science are problems for the historian of science. This facet is equally evident in Gustave Milhaud's *Leçons sur les origines de la science grecque* (Paris: F. Alcan, 1893) and even in the much later work of five impressive volumes, *La Science dans l'antiquité* (Paris: La Renaissance du Livre, 1930–48), by Abel Rey. Only a positivist, utilitarian view of science can justify Rey's astonishing statement that 'the documentary fragments which all come to us from the Orient and Egypt, evoke a first great epoch of science. That it is truly a science according to the *actual* definition of the word, admits no doubt in our eyes' (1:431, *italics* added). That Rey is branded for his 'irrationalist spiritualism' by P. Redondi ('Introductory Notes on Epistemology and the History of Science in France', *Scientia* 110 [1975]:176) is just as incomprehensible as the fact that Redondi's sole grievance against the positivist tradition is that it isolated French historians of science from the interest that prevailed from the 1930s on, under the influence of the Vienna Circle, in questions of conceptual analysis and logic. Redondi fails to see that Duhem's originality was largely due to his openness (dictated not by his 'positivism' but by his Catholicism) to the metaphysical perspectives of intellectual history, perspectives that were most conducive to sensing the importance of questions relating to the origin of science. That there is lasting instructiveness in E. Meyerson's interpretation of scientific history has also much to do with the fact that he had not been a prisoner of the 'metaphysics' of positivism.

39. The crudity of that materialism is made fairly evident in *Scientific Materialism in Nineteenth-Century Germany* (Boston: D. Reidel, 1977) by F. Gregory. As to the historiography of science in late nineteenth-century Germany, its best products were monographs, such as H. Kopp's *Die Entwicklung der Chemie in der neueren Zeit* (Munich: R. Oldenbourg, 1873) and K. A. von Zittel's *Geschichte der Geologie und Paläontologie bis Ende des 19. Jahrhunderts* (Munich: R. Oldenbourg, 1899), that dealt with branches of science neglected in Whewell's *History of the Inductive Sciences*. The histories of physics by F. Rosenberger (*Die Geschichte der Physik*, 1882; reprinted, Hildesheim: Georg Olms, 1965) and by A. Heller (*Geschichte der Physik von Aristoteles bis auf die neueste Zeit*, 1882; reprinted, Wiesbaden: Martin Sändig, 1965) represented improvement only concerning some particulars over Whewell's *History*. About F. Dannemann's *Die Naturwissenschaften in ihrer Entwicklung und in ihrem Zusammenhange* (Leipzig: Wilhelm Engelmann, 1910–13), 'the only modern complete treatise on the history of science' according to G. Sarton (see his *The Study of the History of Science* [Cambridge: Harvard University Press, 1936], p. 63) was found by him 'elementary and imperfect'. Yet Sarton carefully reviewed all its four volumes in *Isis*, a fact strangely conflicting with his silence on Duhem's *Système du monde* except its first volume.

40. The abuse in question was Blanqui's view of comets as holding the key to the explanation of the physical universe and Nietzsche's deliberate oversight of the question of entropy as he tried to graft the idea of eternal returns on the principle of the conservation of energy. For details and documentation, see my *Science and Creation*, pp. 315–16 and 323.

41. Blanqui, who for a while was Marx's competitor for the leadership of the International, is treated, as heretics often are, with silence in Marxist realms. The works of Gustave Le Bon, a physician turned sociologist and philosopher of science, who around the turn of the century endorsed Blanqui's advocacy of eternal recurrences as a fundamental verity, are deservedly forgotten. As examples of very unconvincing efforts to put Nietzsche's science in acceptable light, see G. Batault, 'L'Hypothèse du retour éternel devant la science moderne', *Revue philosophique* 57 (1904): 158–67, and A. Mittasch, *Friedrich Nietzsche als Naturphilosoph* (Stuttgart: Alfred Kröner, 1952).

42. Yet Nietzsche, for instance, most emphatically stated that the moment of scientific truth coincided with the rejection of the doctrine of creation. See aph. 5, in his 'Eternal Recurrence' in *The Complete Works of Friedrich Nietzsche*, edited by Dr. Oscar Levy (Edinburgh: T. N. Foulis, 1909–11), vol. XVI, p. 239.

43. 'The Poverty of Historicism', *Economica* 11 (1944): 86–103, 119–37, and 12 (1945): 68–89.

44. Also evident in Popper's disparaging remarks about the Middle Ages in his *The Open Society and Its Enemies* (Princeton University Press, 1950), pp. 221–2 and 640, and in his systematic oversight of the whole Christian philosophical tradition, an oversight which usually makes him jump from the pre-Socratics to Descartes, although in his latest book, *The Self and Its Brain: An Argument for Interactionism* (Berlin: Springer International, 1977), written jointly with J. C. Eccles, he advances as far as Aristotle in his coverage of pre-Cartesian philosophy.

45. See D. E. Lee and R. N. Beck, 'The Meaning of "Historicism"', *American Historical Review* 59 (1954): 574–5. Those aware of Popper's declaration, 'Science was invented once. It was suppressed by Christianity, and it was only reinvented or, rather, recovered, with the rebirth of Platonism in the Renaissance' (a declaration made at the C. H. Bohringer Sohn Symposium, Kronberg, Taunus, 16–17 May, 1974; see *The Creative Process in Science and Medicine*, ed. H. A. Krebs [Amsterdam: Excerpta Medica, 1975], p. 126), will not expect him to discover the many centuries of Christian philosophy and the vast amount of modern scholarly research on it.

46. The phrase is from the concluding part of the speech Dilthey gave on the occasion of his seventieth birthday in 1903. See *Gesammelte Schriften* (Stuttgart: B. G. Teubner, 1914–74), 5:9. Dilthey failed to realize that nothing short of a complete anarchy of relativism is in store for a scholar who pursued single-mindedly, as he did, a lifelong Kantian criticism of the Hegelian idea of historical 'Vernunft'.

47. His classic *The Idea of History* (Oxford: Clarendon Press, 1946) comes to a close with an emphasis on the difference between scientific and social progress, or rather on the difference between the logic operative in each.

48. Easily the most thematic articulation of that error is P. Frank's *Relativity: A Richer Truth* (Boston: Beacon Press, 1950).

49. *The Structure of Scientific Revolutions* (Chicago: University of Chicago Press, 1962), p. 170.

50. London: Thames and Hudson, 1976, p. 24.

51. His *The Quest for Absolute Zero: The Meaning of Low Temperature Physics* (London: Weidenfeld and Nicolson, 1966) and *The World of Walther Nernst: The Rise and Fall of German Science* (London: Macmillan, 1973) fall short of basic standards required nowadays in scholarly works in the history of science. As to his *The Riddle of the Pyramids* (London: Thames and Hudson, 1974), it shows the virulent apriorism of a Marxist thinker for whom historical account and explanation must be cast in terms of sociological reductionism. It is

stated in that book that 'the foremost duty of a scientist is to stick to the facts' (p. 177). What Mendelssohn clearly fails to understand is that there are many more facts to history, including the history of science, than a scientist as such can dream of.

52. *Science and Western Domination*, p. 49.

53. Ibid., p. 18.

54. Those days were, it is well to remember, still around a generation ago when many scientists working on the atomic bomb felt exactly that way as admitted by one of them, James Franck: 'So we took the easiest way out and hid in our ivory tower. We felt that neither the good nor the evil applications were our responsibility.' See his essay, 'The Social Task of the Scientist', *Bulletin of the Atomic Scientists* 3 (1947): 70.

55. H. Melville, *Moby Dick, or the Whale*, edited by L. S. Mansfield and H. P. Vincent (New York: Hendricks House, 1952), p. 185.

56. In a letter to P. Bayle; see *Leibniz Selections*, edited by P. P. Wiener (New York: Charles Scribner's Sons, 1951), p. 70.

57. Ibid.

58. See Leibniz's 'Discourse on Metaphysics' (1686), ibid., pp. 322–3. But he found the work 'a little too long'. A year later, in 1687, he merely stated that the *Phaedo* 'should be read in entirety'; ibid., pp. 326 and 69.

59. A harsh but not altogether unjust evaluation of it by E. T. Whittaker in his *From Euclid to Eddington: A Study of Conceptions of the External World* (Cambridge: University Press, 1949), p. 65.

60. Its most telling result was the scientific methodology embodied in the motto, 'to save the phenomena', standing for strict mathematical formalism severed from ontological considerations of physical reality. Thus the Socratic programme, 'to save the purpose', could still be honoured through a 'physical' explanation which was markedly organismic or biological. Yet such an explanation could easily become subjectivist and arbitrary, as shown by the case of Ptolemy, a mathematical formalist in his *Almagest*, but a 'realist' in his *Planetary hypotheses*, where he explained the smooth motion of planets with a reference to the instinctive harmony guiding the movements of a group of dancers. The subjectivism germane to such a 'scientific realism' may help explain the fact, exposed with a vengeance by Robert R. Newton in his *The Crime of Ptolemy* (Baltimore: Johns Hopkins Press, 1977), of Ptolemy's blithe manipulation in his *Almagest* of data of observations!

61. The expression is a chapter title in *Four Stages of Greek Religion* by G. Murray (New York: Columbia University Press, 1912), p. 103. The Greeks' loss of confidence from the fourth century on is

repeatedly noted in *The Greek Experience* (1957; London: Cardinal Editions, 1973, pp. 33, 206, 216, and 220) by C. M. Bowra, who saw its source in the disastrous outcome for Athens of the Peloponnesian Wars. This explanation is, however, difficult to reconcile with the flourishing of Greek science and philosophy during the fourth century.

62. The result was a subtle relapse into animism and its subjectivism, a fact well illustrated by Kant's persistently inconsistent utterances on teleology, the topic of *Kant's Concept of Teleology* by J. D. McFarland (Edinburgh: University of Edinburgh Press, 1970).

63. Darwin was particularly pleased when Asa Gray wrote in his article 'Charles Darwin' (*Nature*, June 4, 1874, p. 81) that in view of Darwin's great service to Natural Science '. . . instead of Morphology versus Teleology we shall have Morphology wedded to Teleology'. For Darwin's approval of this see *The Life and Letters of Charles Darwin*, edited by F. Darwin (London: J. Murray, 1888), 3:189.

64. The fallacies of this position are ably set forth by R. V. Young, Jr., 'Christianity and Ecology', *National Review* 26 (1974):1454–8, 1477, and 1479.

65. 'The Historical Roots of Our Ecologic Crisis', *Science* 155 (1967):1203–1207. The importance of White's article is largely due to its widespread publicity enhanced by reprints in several anthologies relating to the ecological impact of technology and the ethical problems raised thereby. The remark, that 'historians and theologians have long pointed out that technological progress may have more than an accidental relationship to biblical faith', is the sole and hardly illuminating reference to the role of that faith in the rise of science in the spirited criticism of White's position by R. L. Shinn, 'Science and Ethical Decision: Some New Issues', in *Earth Might Be Fair: Reflections on Ethics, Religion, and Ecology*, edited by I. G. Barbour (Englewood Cliffs, N.J.: Prentice-Hall, 1972), pp. 123–45; for quotation, see p. 140.

66. 'The Faith of a Humanist', in *The History of Science and the New Humanism* (Bloomington: Indiana University Press, 1962), pp. x–xi. In his article quoted in the foregoing note, White also extols Saint Francis' attitude as opposed to a 'dominating' type of Christianity. The true logic of insisting on such a dichotomy is amply revealed by the fact that White urges as a solution not the emulation of Francis' Christianity but of that Buddhism which he had already credited with the stimulation of the Christian medieval West concerning science and technology. See note 38 to Lecture One.

67. *The Impact of Science on Society* (New York: Simon and Schuster, 1953), p. 92.

68. Huxley's letter of September 23, 1860, to the Rev. Charles Kingsley, who upon the death of Huxley's seven-year-old son tried to induce Huxley to give some consideration to Christian hope in resurrection. See L. Huxley, *The Life and Letters of Thomas Henry Huxley* (London: Macmillan, 1900), vol. 1, p. 219.

INDEX OF NAMES